Fire Alarm Log Book

COMPANY DETAILS

COMPANY NAME

ADDRESS

E-MAIL ADDRESS

WEBSITE

PHONE   **FAX**

EMERGENCY CONTACT PERSON

PHONE   **FAX**

LOG BOOK DETAILS

CONTINUED FROM LOG BOOK

LOG START DATE

CONTINUED TO LOG BOOK

LOG END DATE

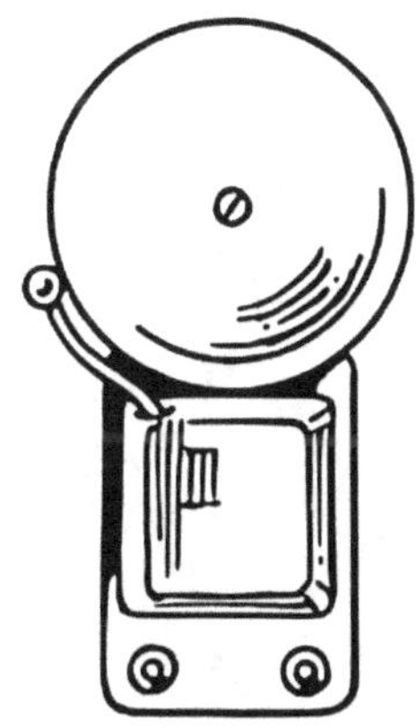

Fire Alarm Log Book

YEAR   MONTH

DATE   TIME

LOCATION   CALL NO.

CHECKS DONE

ACTION REQUIRED

ACTION TAKEN

DATE ACTION WAS LOGGED   LOGGED BY

DATE ACTION WAS CLOSED   CLOSED BY

NOTES

Fire Alarm Log Book

YEAR   MONTH

DATE   TIME

LOCATION   CALL NO.

CHECKS DONE

ACTION REQUIRED

ACTION TAKEN

DATE ACTION WAS LOGGED   LOGGED BY

DATE ACTION WAS CLOSED   CLOSED BY

NOTES

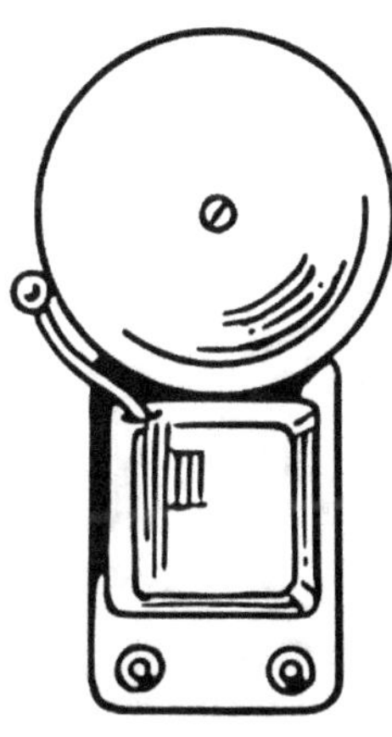

Fire Alarm Log Book

YEAR   MONTH

DATE   TIME

LOCATION   CALL NO.

CHECKS DONE

ACTION REQUIRED

ACTION TAKEN

DATE ACTION WAS LOGGED   LOGGED BY

DATE ACTION WAS CLOSED   CLOSED BY

NOTES

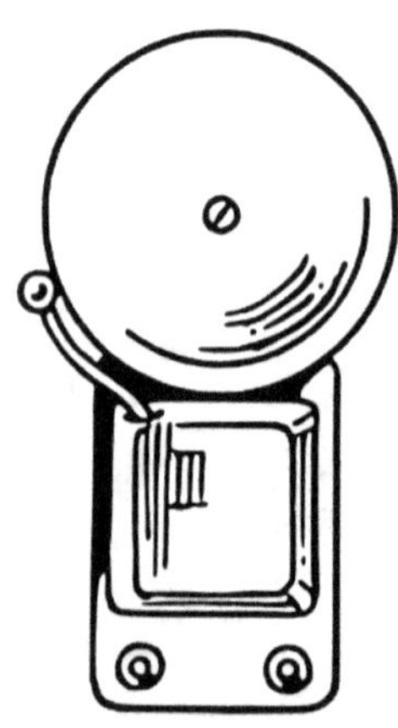

Fire Alarm Log Book

YEAR   MONTH

DATE   TIME

LOCATION   CALL NO.

CHECKS DONE

ACTION REQUIRED

ACTION TAKEN

DATE ACTION WAS LOGGED   LOGGED BY

DATE ACTION WAS CLOSED   CLOSED BY

NOTES

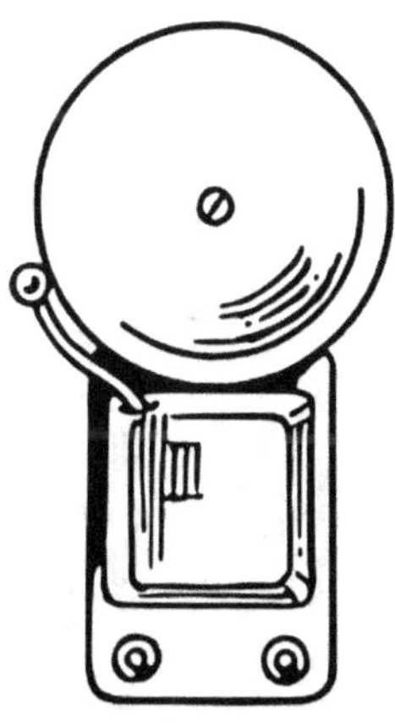

Fire Alarm Log Book 

YEAR

MONTH

DATE

TIME

LOCATION

CALL NO.

CHECKS DONE

ACTION REQUIRED

ACTION TAKEN

DATE ACTION WAS LOGGED

LOGGED BY

DATE ACTION WAS CLOSED

CLOSED BY

NOTES

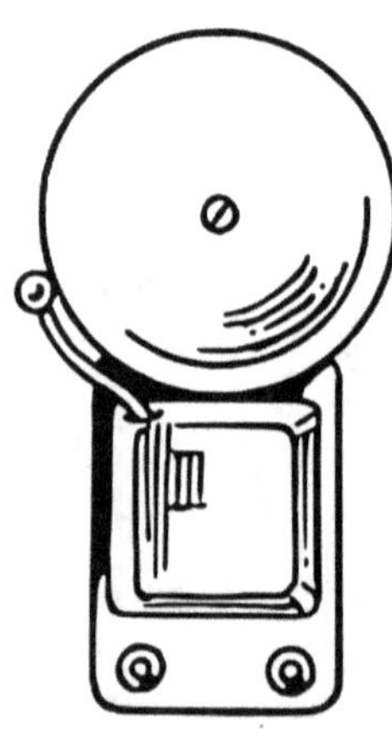

Fire Alarm Log Book

YEAR   **MONTH**

DATE   **TIME**

LOCATION   **CALL NO.**

CHECKS DONE

ACTION REQUIRED

ACTION TAKEN

DATE ACTION WAS LOGGED   **LOGGED BY**

DATE ACTION WAS CLOSED   **CLOSED BY**

NOTES

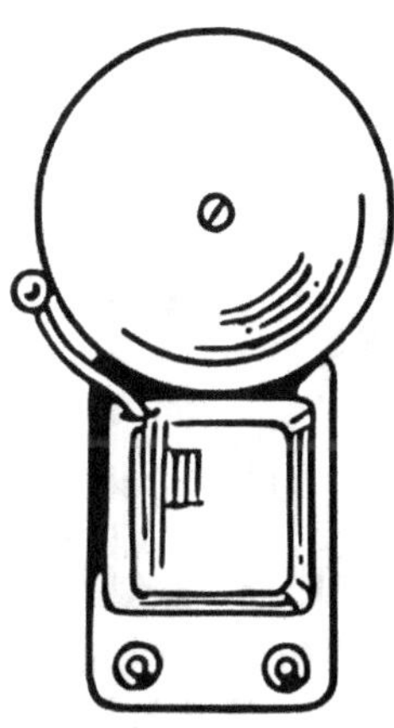

Fire Alarm Log Book

YEAR	**MONTH**
DATE	**TIME**
LOCATION	**CALL NO.**

CHECKS DONE

ACTION REQUIRED

ACTION TAKEN

DATE ACTION WAS LOGGED	**LOGGED BY**
DATE ACTION WAS CLOSED	**CLOSED BY**

NOTES

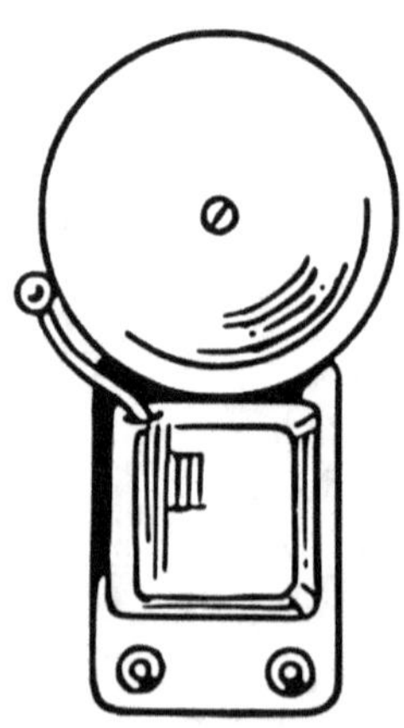

Fire Alarm Log Book

YEAR   MONTH

DATE   TIME

LOCATION   CALL NO.

CHECKS DONE

ACTION REQUIRED

ACTION TAKEN

DATE ACTION WAS LOGGED   LOGGED BY

DATE ACTION WAS CLOSED   CLOSED BY

NOTES

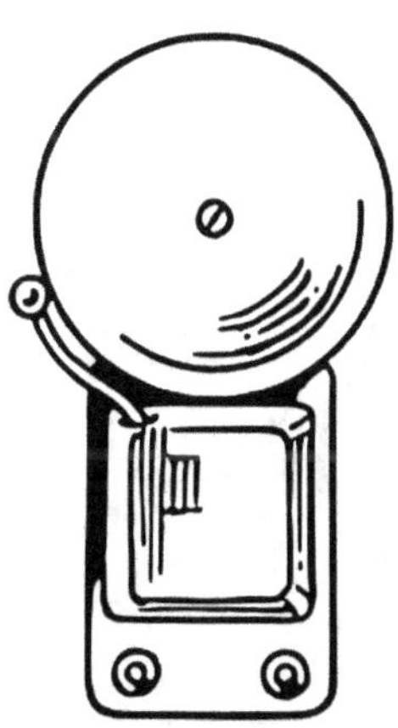

Fire Alarm Log Book

YEAR   MONTH

DATE   TIME

LOCATION   CALL NO.

CHECKS DONE

ACTION REQUIRED

ACTION TAKEN

DATE ACTION WAS LOGGED   LOGGED BY

DATE ACTION WAS CLOSED   CLOSED BY

NOTES

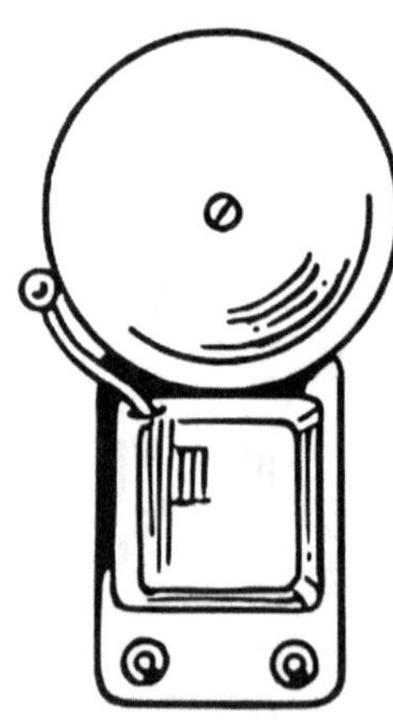

Fire Alarm Log Book

YEAR   MONTH

DATE   TIME

LOCATION   CALL NO.

CHECKS DONE

ACTION REQUIRED

ACTION TAKEN

DATE ACTION WAS LOGGED   LOGGED BY

DATE ACTION WAS CLOSED   CLOSED BY

NOTES

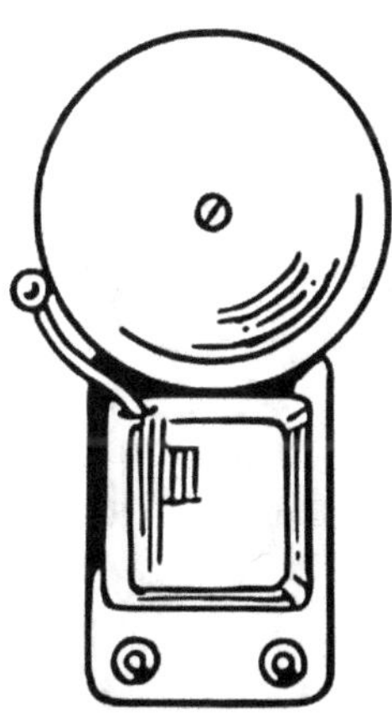

Fire Alarm Log Book

YEAR   MONTH

DATE   TIME

LOCATION   CALL NO.

CHECKS DONE

ACTION REQUIRED

ACTION TAKEN

DATE ACTION WAS LOGGED   LOGGED BY

DATE ACTION WAS CLOSED   CLOSED BY

NOTES

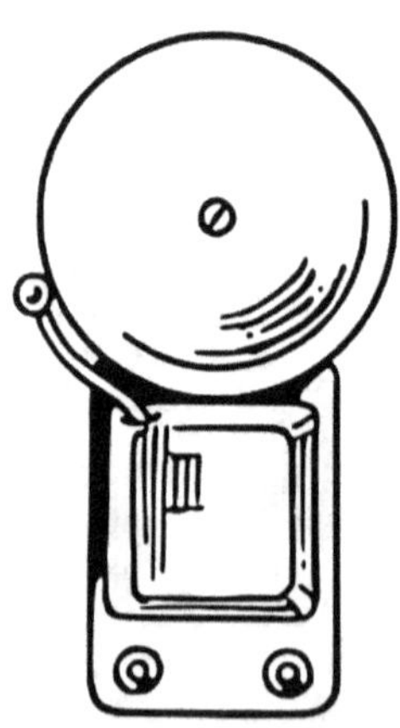

Fire Alarm Log Book

YEAR   MONTH

DATE   TIME

LOCATION   CALL NO.

CHECKS DONE

ACTION REQUIRED

ACTION TAKEN

DATE ACTION WAS LOGGED   LOGGED BY

DATE ACTION WAS CLOSED   CLOSED BY

NOTES

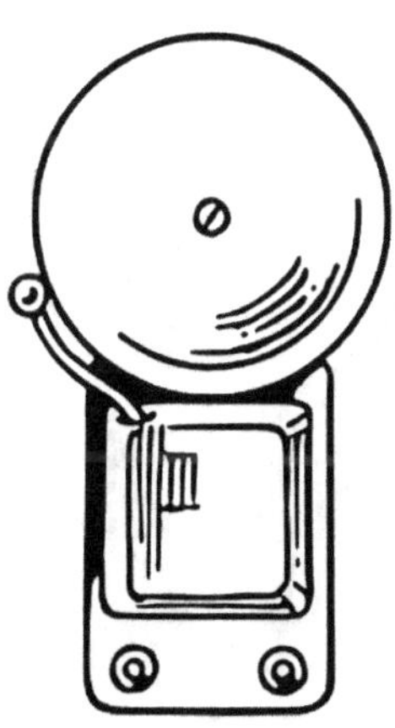

Fire Alarm Log Book

YEAR	**MONTH**
DATE	**TIME**
LOCATION	**CALL NO.**

CHECKS DONE

ACTION REQUIRED

ACTION TAKEN

DATE ACTION WAS LOGGED	**LOGGED BY**
DATE ACTION WAS CLOSED	**CLOSED BY**

NOTES

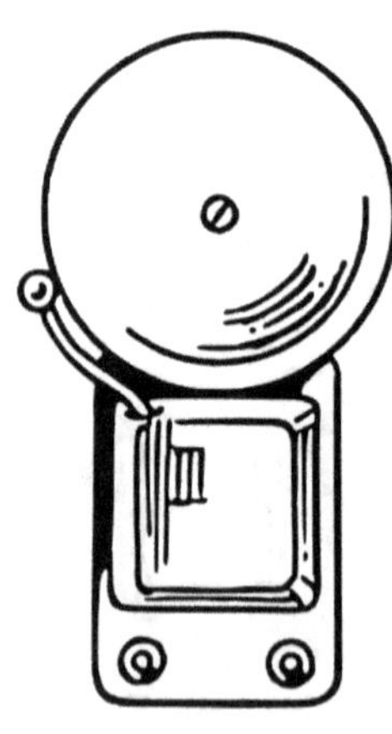

Fire Alarm Log Book

YEAR

MONTH

DATE

TIME

LOCATION

CALL NO.

CHECKS DONE

ACTION REQUIRED

ACTION TAKEN

DATE ACTION WAS LOGGED

LOGGED BY

DATE ACTION WAS CLOSED

CLOSED BY

NOTES

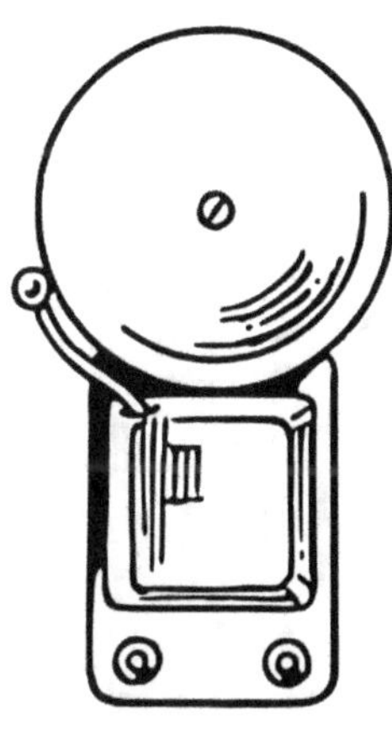

Fire Alarm Log Book

YEAR   MONTH

DATE   TIME

LOCATION   CALL NO.

CHECKS DONE

ACTION REQUIRED

ACTION TAKEN

DATE ACTION WAS LOGGED   LOGGED BY

DATE ACTION WAS CLOSED   CLOSED BY

NOTES

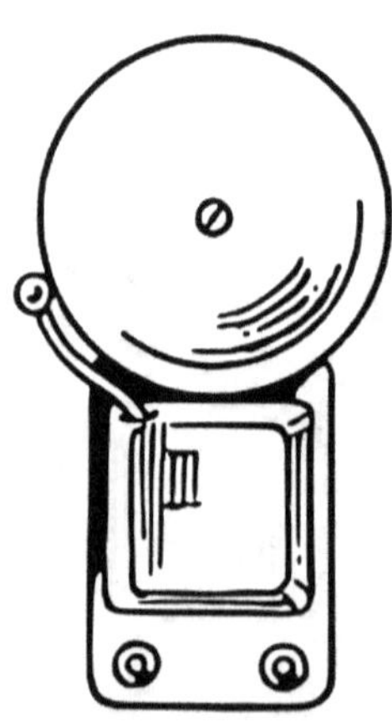

Fire Alarm Log Book

YEAR   MONTH

DATE   TIME

LOCATION   CALL NO.

CHECKS DONE

ACTION REQUIRED

ACTION TAKEN

DATE ACTION WAS LOGGED   LOGGED BY

DATE ACTION WAS CLOSED   CLOSED BY

NOTES

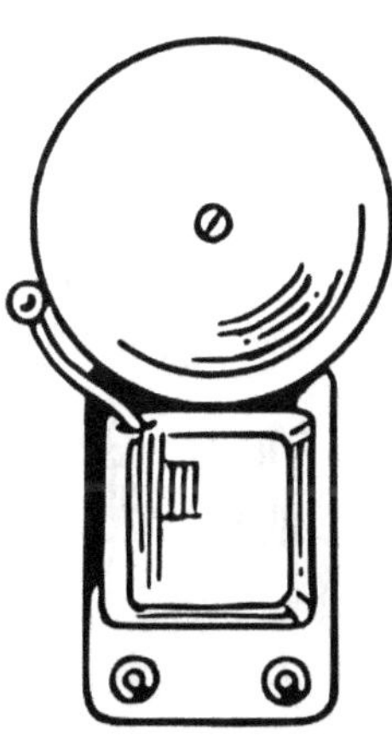

Fire Alarm Log Book

YEAR   MONTH

DATE   TIME

LOCATION   CALL NO.

CHECKS DONE

ACTION REQUIRED

ACTION TAKEN

DATE ACTION WAS LOGGED   LOGGED BY

DATE ACTION WAS CLOSED   CLOSED BY

NOTES

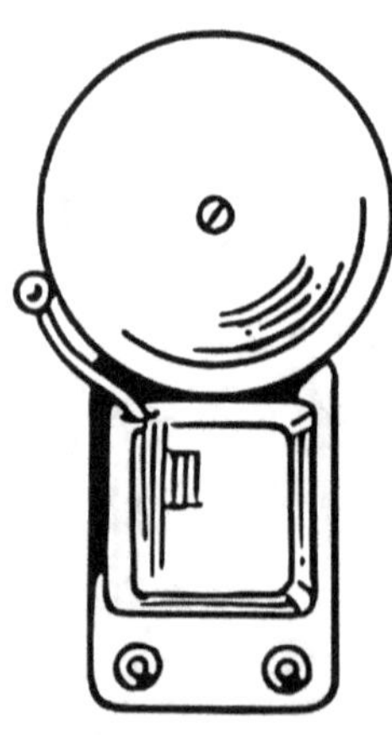

Fire Alarm Log Book

YEAR .... MONTH ....

DATE .... TIME ....

LOCATION .... CALL NO. ....

CHECKS DONE

ACTION REQUIRED

ACTION TAKEN

DATE ACTION WAS LOGGED .... LOGGED BY ....

DATE ACTION WAS CLOSED .... CLOSED BY ....

NOTES

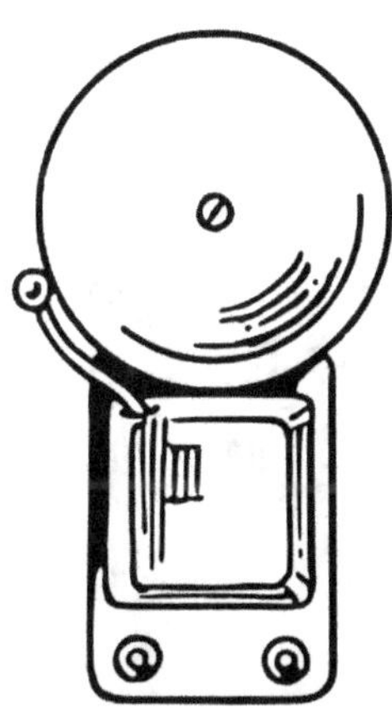

Fire Alarm Log Book

YEAR   MONTH

DATE   TIME

LOCATION   CALL NO.

CHECKS DONE

ACTION REQUIRED

ACTION TAKEN

DATE ACTION WAS LOGGED   LOGGED BY

DATE ACTION WAS CLOSED   CLOSED BY

NOTES

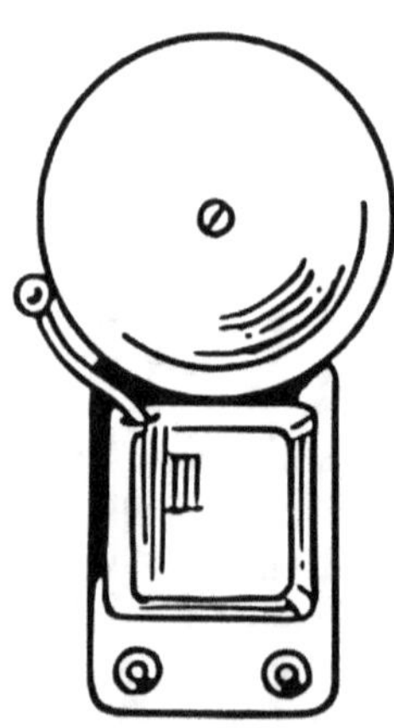

Fire Alarm Log Book

YEAR   MONTH

DATE   TIME

LOCATION   CALL NO.

CHECKS DONE

ACTION REQUIRED

ACTION TAKEN

DATE ACTION WAS LOGGED   LOGGED BY

DATE ACTION WAS CLOSED   CLOSED BY

NOTES

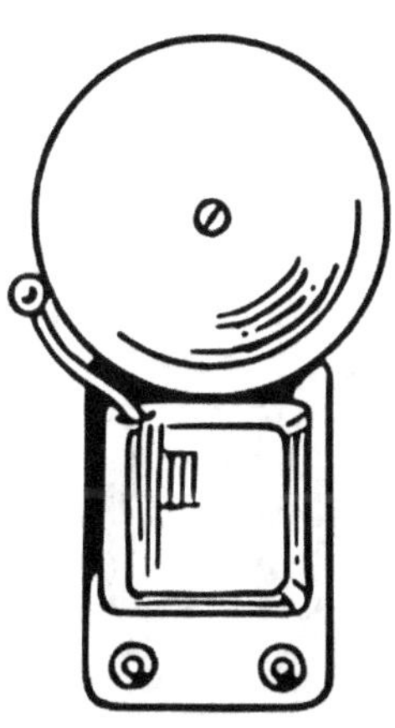

Fire Alarm Log Book

YEAR   MONTH

DATE   TIME

LOCATION   CALL NO.

CHECKS DONE

ACTION REQUIRED

ACTION TAKEN

DATE ACTION WAS LOGGED   LOGGED BY

DATE ACTION WAS CLOSED   CLOSED BY

NOTES

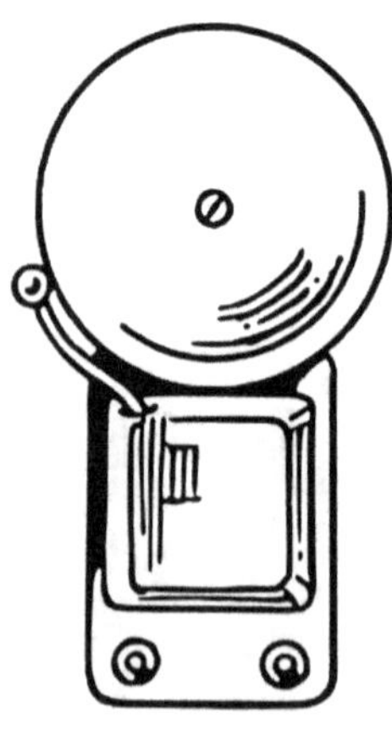

Fire Alarm Log Book

YEAR

MONTH

DATE

TIME

LOCATION

CALL NO.

CHECKS DONE

ACTION REQUIRED

ACTION TAKEN

DATE ACTION WAS LOGGED

LOGGED BY

DATE ACTION WAS CLOSED

CLOSED BY

NOTES

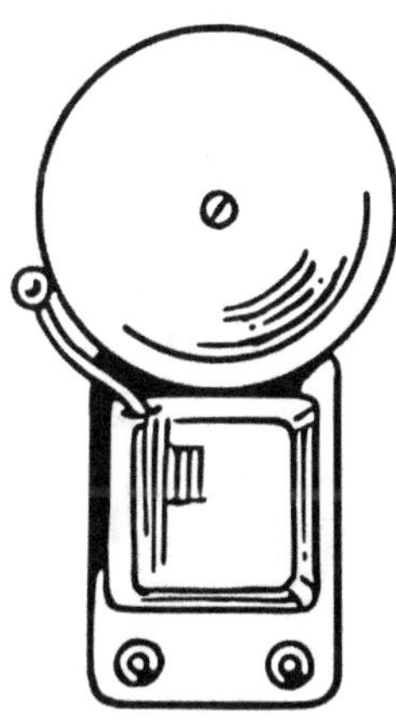

Fire Alarm Log Book 

YEAR   MONTH

DATE   TIME

LOCATION   CALL NO.

CHECKS DONE

ACTION REQUIRED

ACTION TAKEN

DATE ACTION WAS LOGGED   LOGGED BY

DATE ACTION WAS CLOSED   CLOSED BY

NOTES

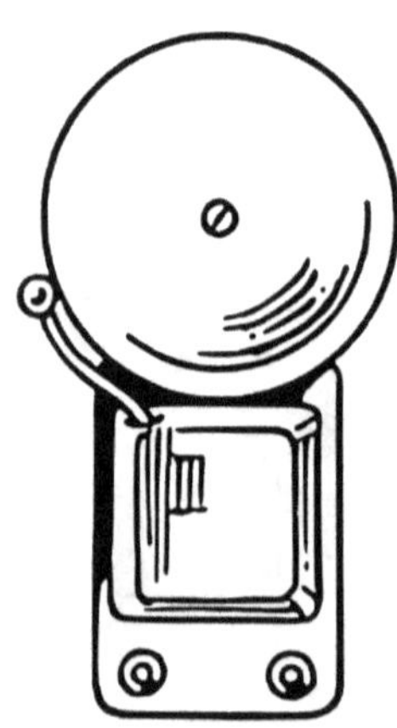

Fire Alarm Log Book

YEAR   MONTH

DATE   TIME

LOCATION   CALL NO.

CHECKS DONE

ACTION REQUIRED

ACTION TAKEN

DATE ACTION WAS LOGGED   LOGGED BY

DATE ACTION WAS CLOSED   CLOSED BY

NOTES

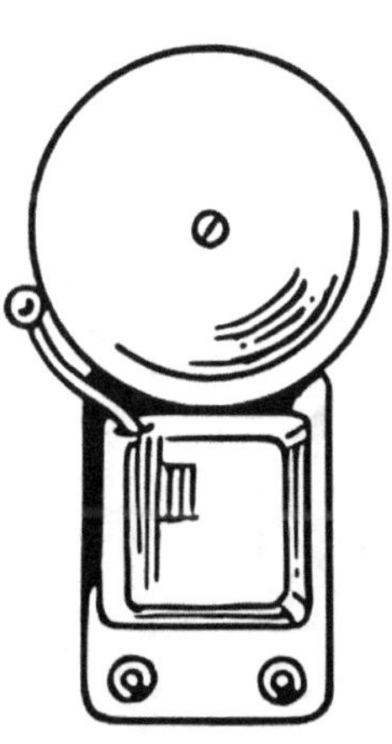

Fire Alarm Log Book

YEAR   MONTH

DATE   TIME

LOCATION   CALL NO.

CHECKS DONE

ACTION REQUIRED

ACTION TAKEN

DATE ACTION WAS LOGGED   LOGGED BY

DATE ACTION WAS CLOSED   CLOSED BY

NOTES

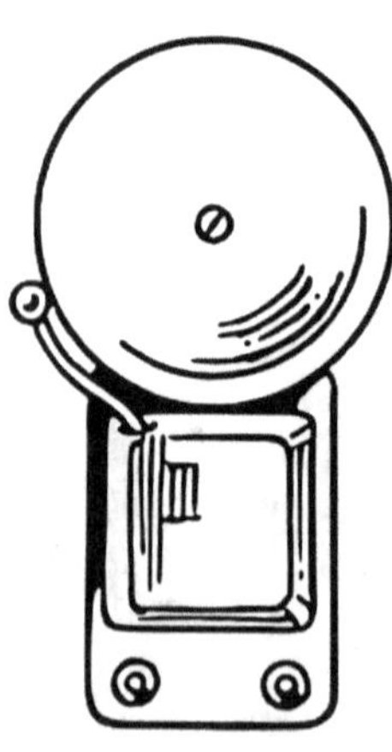

Fire Alarm Log Book

YEAR

MONTH

DATE

TIME

LOCATION

CALL NO.

CHECKS DONE

ACTION REQUIRED

ACTION TAKEN

DATE ACTION WAS LOGGED

LOGGED BY

DATE ACTION WAS CLOSED

CLOSED BY

NOTES

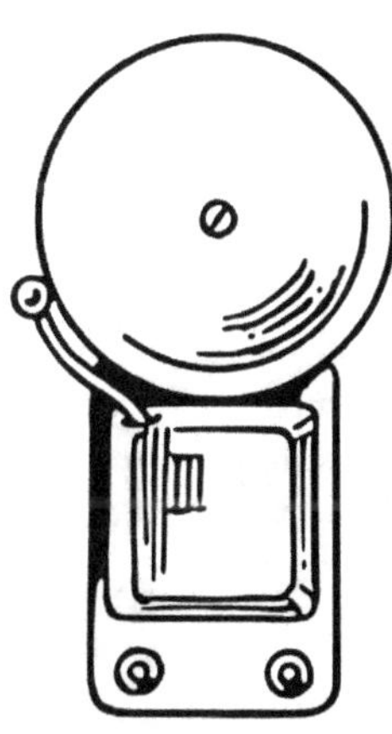

Fire Alarm Log Book

YEAR　　　　　　　　　　　MONTH

DATE　　　　　　　　　　　TIME

LOCATION　　　　　　　　　CALL NO.

CHECKS DONE

ACTION REQUIRED

ACTION TAKEN

DATE ACTION WAS LOGGED　　　　　　　　　　LOGGED BY

DATE ACTION WAS CLOSED　　　　　　　　　　CLOSED BY

NOTES

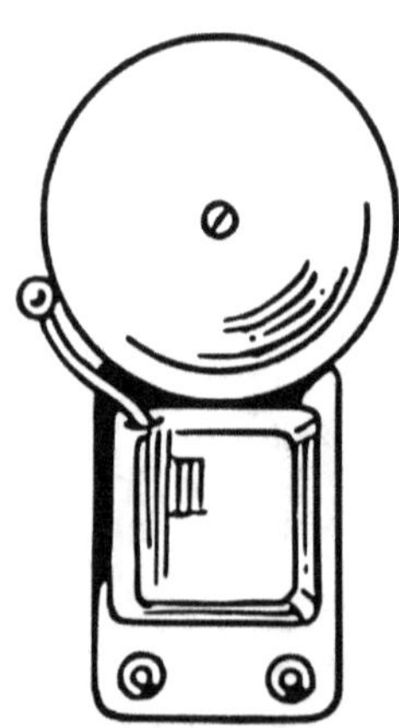

Fire Alarm Log Book

YEAR   MONTH

DATE   TIME

LOCATION   CALL NO.

CHECKS DONE

ACTION REQUIRED

ACTION TAKEN

DATE ACTION WAS LOGGED   LOGGED BY

DATE ACTION WAS CLOSED   CLOSED BY

NOTES

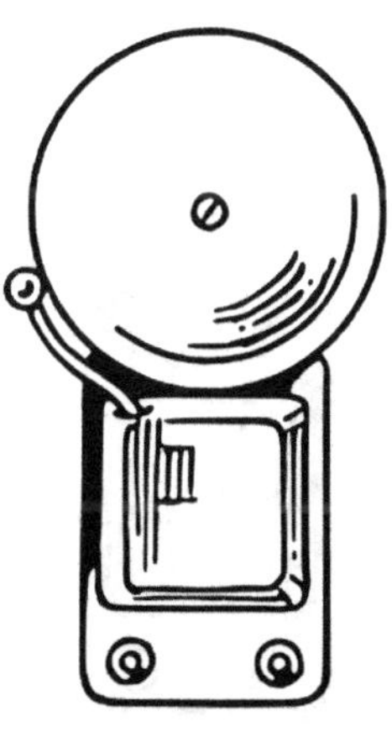

Fire Alarm Log Book

YEAR   MONTH

DATE   TIME

LOCATION   CALL NO.

CHECKS DONE

ACTION REQUIRED

ACTION TAKEN

DATE ACTION WAS LOGGED   LOGGED BY

DATE ACTION WAS CLOSED   CLOSED BY

NOTES

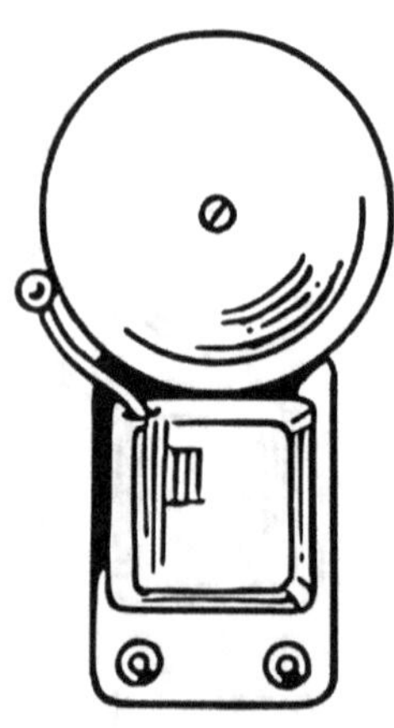

Fire Alarm Log Book 

YEAR

MONTH

DATE

TIME

LOCATION

CALL NO.

CHECKS DONE

ACTION REQUIRED

ACTION TAKEN

DATE ACTION WAS LOGGED

LOGGED BY

DATE ACTION WAS CLOSED

CLOSED BY

NOTES

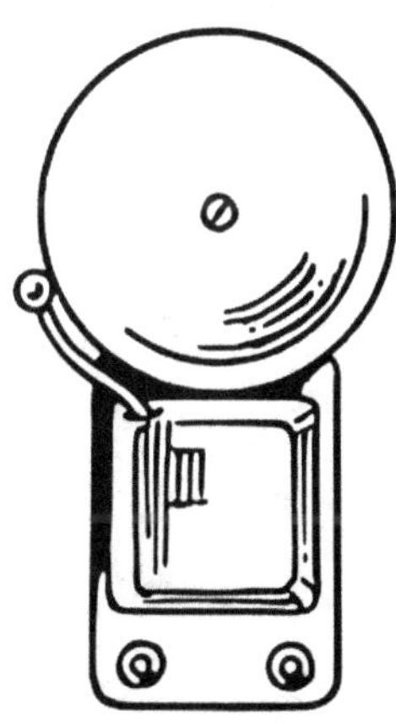

Fire Alarm Log Book

YEAR   MONTH

DATE   TIME

LOCATION   CALL NO.

CHECKS DONE

ACTION REQUIRED

ACTION TAKEN

DATE ACTION WAS LOGGED   LOGGED BY

DATE ACTION WAS CLOSED   CLOSED BY

NOTES

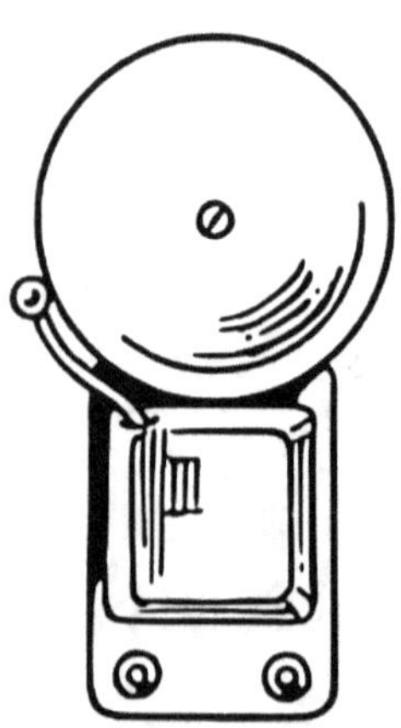

Fire Alarm Log Book

YEAR	**MONTH**
DATE	**TIME**
LOCATION	**CALL NO.**

CHECKS DONE

ACTION REQUIRED

ACTION TAKEN

DATE ACTION WAS LOGGED	**LOGGED BY**
DATE ACTION WAS CLOSED	**CLOSED BY**

NOTES

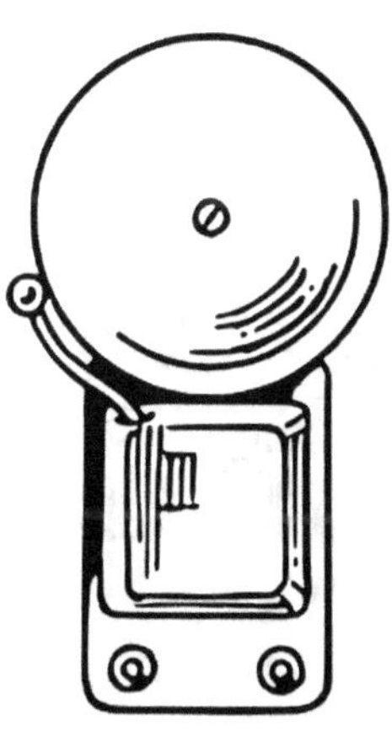

Fire Alarm Log Book

YEAR

MONTH

DATE

TIME

LOCATION

CALL NO.

CHECKS DONE

ACTION REQUIRED

ACTION TAKEN

DATE ACTION WAS LOGGED

LOGGED BY

DATE ACTION WAS CLOSED

CLOSED BY

NOTES

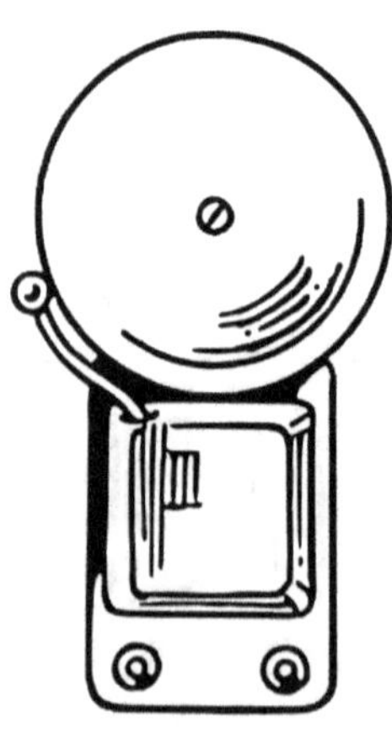

Fire Alarm Log Book

YEAR

MONTH

DATE

TIME

LOCATION

CALL NO.

CHECKS DONE

ACTION REQUIRED

ACTION TAKEN

DATE ACTION WAS LOGGED

LOGGED BY

DATE ACTION WAS CLOSED

CLOSED BY

NOTES

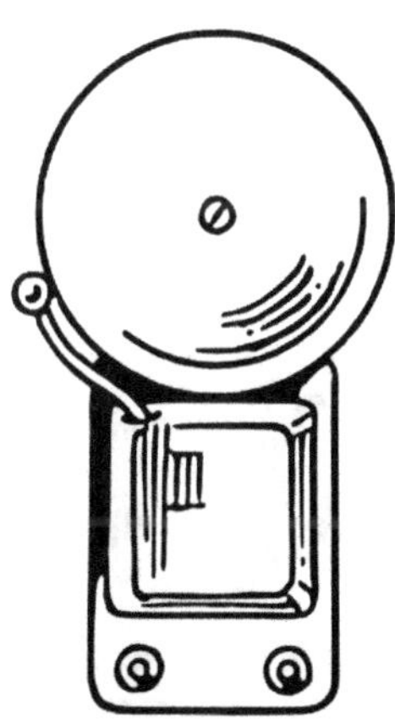

Fire Alarm Log Book

YEAR

MONTH

DATE

TIME

LOCATION

CALL NO.

CHECKS DONE

ACTION REQUIRED

ACTION TAKEN

DATE ACTION WAS LOGGED

LOGGED BY

DATE ACTION WAS CLOSED

CLOSED BY

NOTES

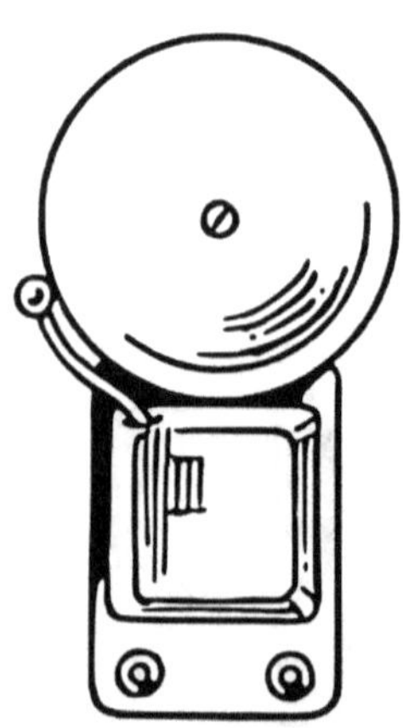

Fire Alarm Log Book

YEAR	**MONTH**
DATE	**TIME**
LOCATION	**CALL NO.**

CHECKS DONE

ACTION REQUIRED

ACTION TAKEN

DATE ACTION WAS LOGGED	**LOGGED BY**
DATE ACTION WAS CLOSED	**CLOSED BY**

NOTES

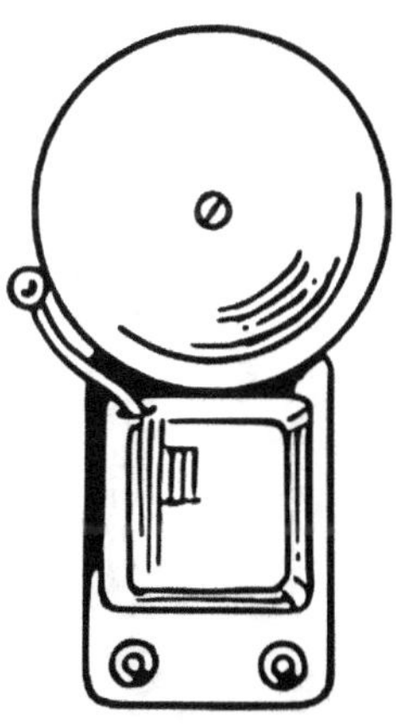

Fire Alarm Log Book

YEAR	**MONTH**
DATE	**TIME**
LOCATION	**CALL NO.**

CHECKS DONE

ACTION REQUIRED

ACTION TAKEN

DATE ACTION WAS LOGGED	**LOGGED BY**
DATE ACTION WAS CLOSED	**CLOSED BY**

NOTES

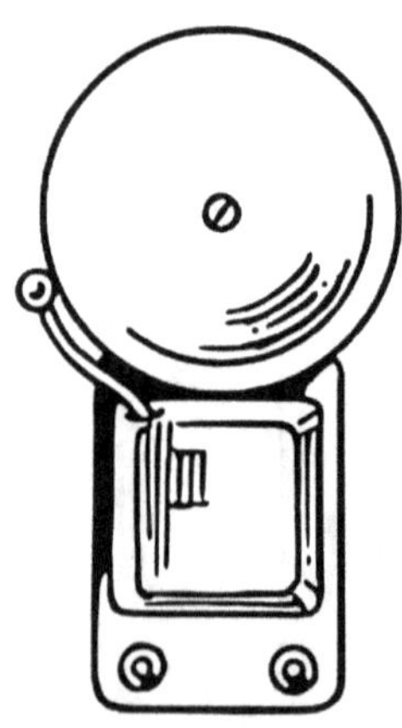

Fire Alarm Log Book

YEAR   MONTH

DATE   TIME

LOCATION   CALL NO.

CHECKS DONE

ACTION REQUIRED

ACTION TAKEN

DATE ACTION WAS LOGGED   LOGGED BY

DATE ACTION WAS CLOSED   CLOSED BY

NOTES

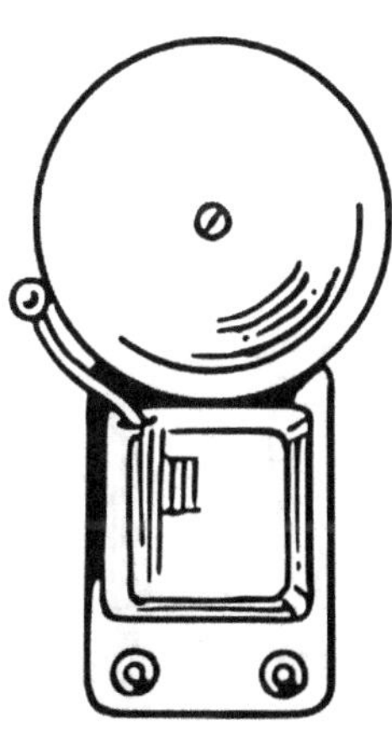

Fire Alarm Log Book

YEAR   MONTH

DATE   TIME

LOCATION   CALL NO.

CHECKS DONE

ACTION REQUIRED

ACTION TAKEN

DATE ACTION WAS LOGGED   LOGGED BY

DATE ACTION WAS CLOSED   CLOSED BY

NOTES

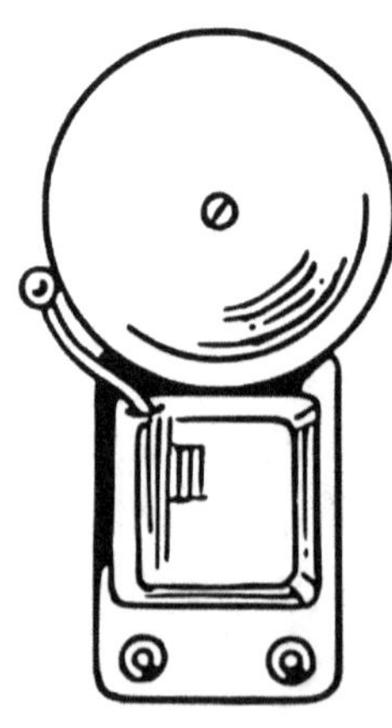

Fire Alarm Log Book

YEAR   MONTH

DATE   TIME

LOCATION   CALL NO.

CHECKS DONE

ACTION REQUIRED

ACTION TAKEN

DATE ACTION WAS LOGGED   LOGGED BY

DATE ACTION WAS CLOSED   CLOSED BY

NOTES

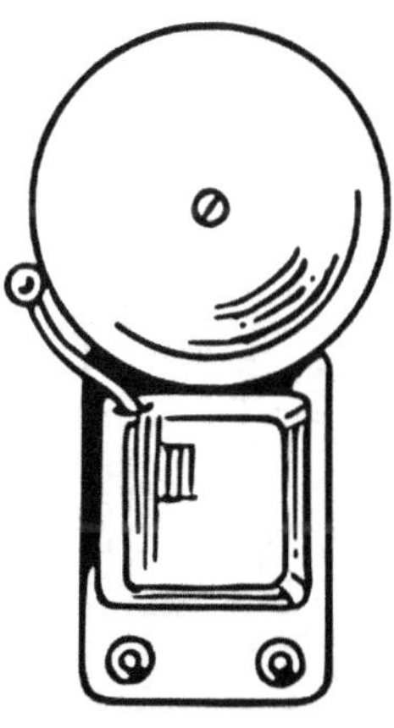

Fire Alarm Log Book

YEAR

MONTH

DATE

TIME

LOCATION

CALL NO.

CHECKS DONE

ACTION REQUIRED

ACTION TAKEN

DATE ACTION WAS LOGGED

LOGGED BY

DATE ACTION WAS CLOSED

CLOSED BY

NOTES

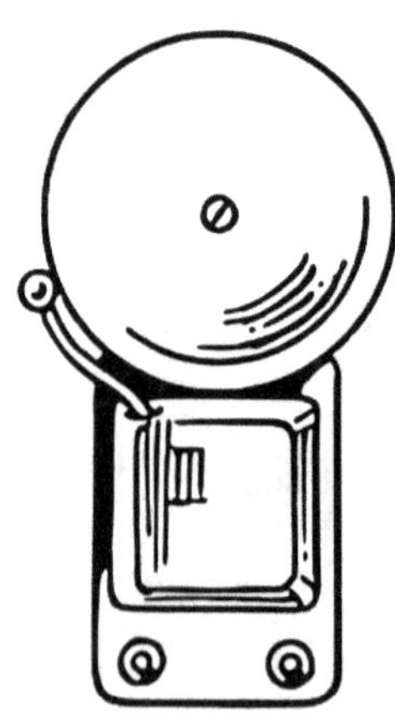

Fire Alarm Log Book

YEAR	**MONTH**
DATE	**TIME**
LOCATION	**CALL NO.**

CHECKS DONE

ACTION REQUIRED

ACTION TAKEN

DATE ACTION WAS LOGGED	**LOGGED BY**
DATE ACTION WAS CLOSED	**CLOSED BY**

NOTES

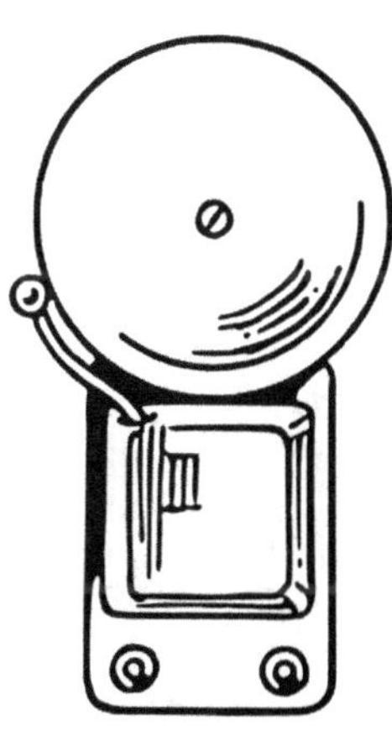

Fire Alarm Log Book

YEAR   MONTH

DATE   TIME

LOCATION   CALL NO.

CHECKS DONE

ACTION REQUIRED

ACTION TAKEN

DATE ACTION WAS LOGGED   LOGGED BY

DATE ACTION WAS CLOSED   CLOSED BY

NOTES

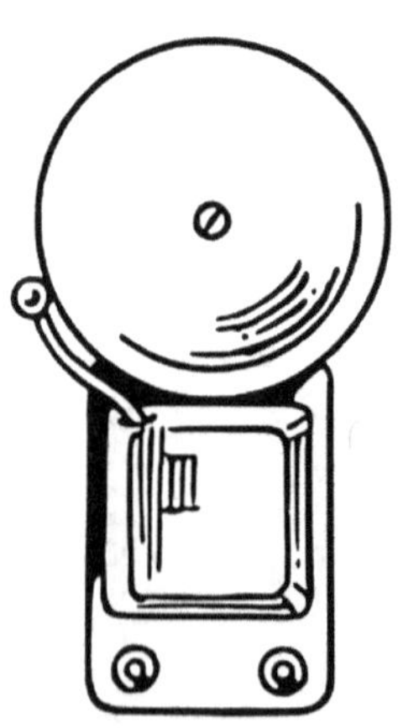

Fire Alarm Log Book

YEAR		**MONTH**	
DATE		**TIME**	
LOCATION		**CALL NO.**	

CHECKS DONE

ACTION REQUIRED

ACTION TAKEN

DATE ACTION WAS LOGGED		**LOGGED BY**	
DATE ACTION WAS CLOSED		**CLOSED BY**	

NOTES

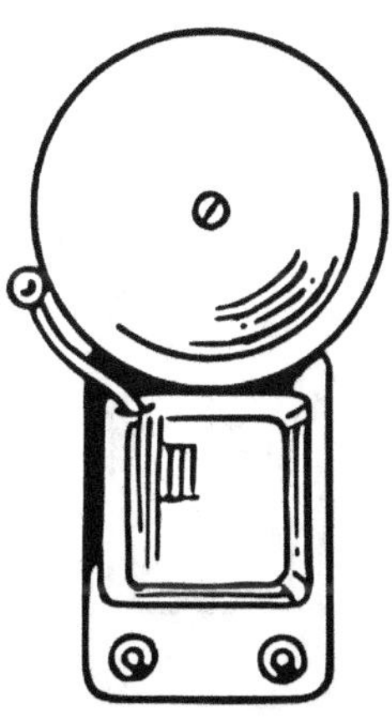

Fire Alarm Log Book

YEAR   MONTH

DATE   TIME

LOCATION   CALL NO.

CHECKS DONE

ACTION REQUIRED

ACTION TAKEN

DATE ACTION WAS LOGGED   LOGGED BY

DATE ACTION WAS CLOSED   CLOSED BY

NOTES

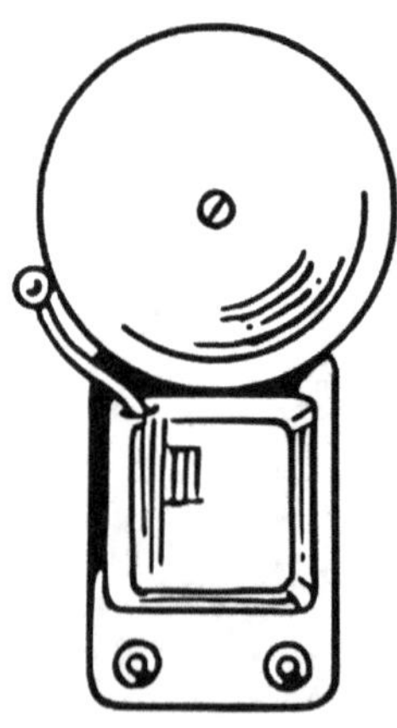

Fire Alarm Log Book

YEAR | MONTH

DATE | TIME

LOCATION | CALL NO.

CHECKS DONE

ACTION REQUIRED

ACTION TAKEN

DATE ACTION WAS LOGGED | LOGGED BY

DATE ACTION WAS CLOSED | CLOSED BY

NOTES

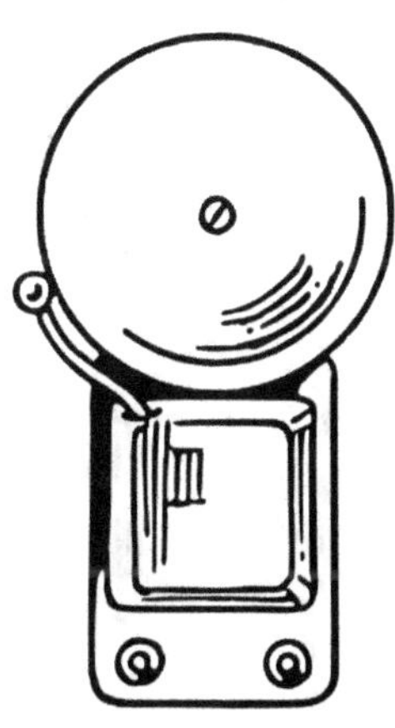

Fire Alarm Log Book 

YEAR   **MONTH**

DATE   **TIME**

LOCATION   **CALL NO.**

CHECKS DONE

ACTION REQUIRED

ACTION TAKEN

DATE ACTION WAS LOGGED   **LOGGED BY**

DATE ACTION WAS CLOSED   **CLOSED BY**

NOTES

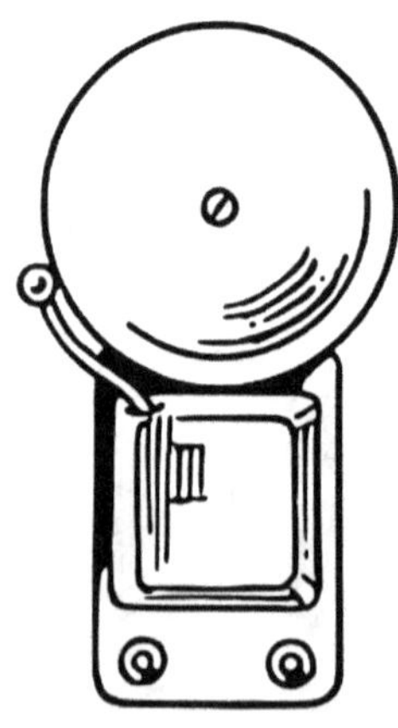

Fire Alarm Log Book

YEAR   MONTH

DATE   TIME

LOCATION   CALL NO.

CHECKS DONE

ACTION REQUIRED

ACTION TAKEN

DATE ACTION WAS LOGGED   LOGGED BY

DATE ACTION WAS CLOSED   CLOSED BY

NOTES

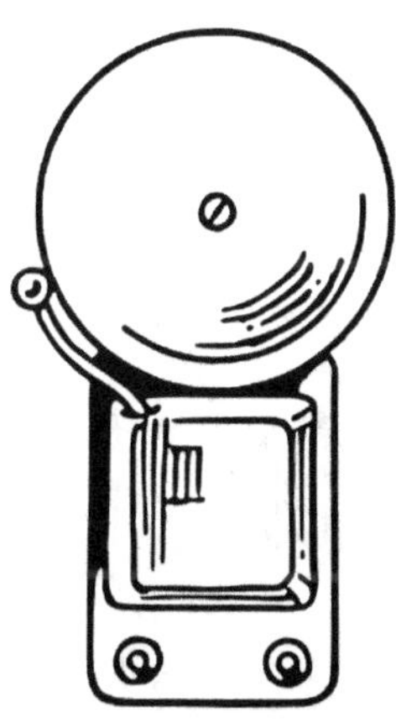

Fire Alarm Log Book

YEAR	**MONTH**
DATE	**TIME**
LOCATION	**CALL NO.**

CHECKS DONE

ACTION REQUIRED

ACTION TAKEN

DATE ACTION WAS LOGGED	**LOGGED BY**
DATE ACTION WAS CLOSED	**CLOSED BY**

NOTES

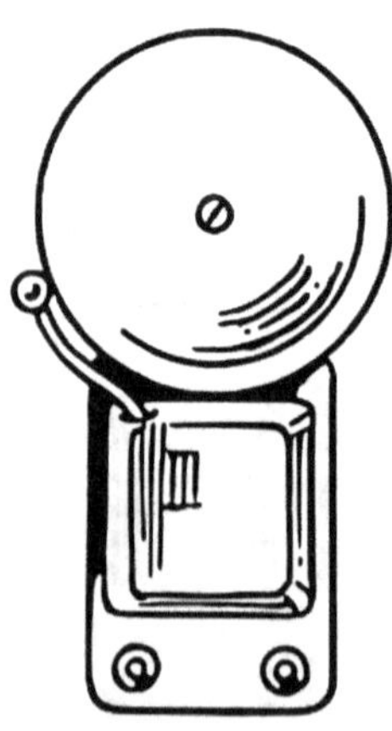

Fire Alarm Log Book

YEAR   MONTH

DATE   TIME

LOCATION   CALL NO.

CHECKS DONE

ACTION REQUIRED

ACTION TAKEN

DATE ACTION WAS LOGGED   LOGGED BY

DATE ACTION WAS CLOSED   CLOSED BY

NOTES

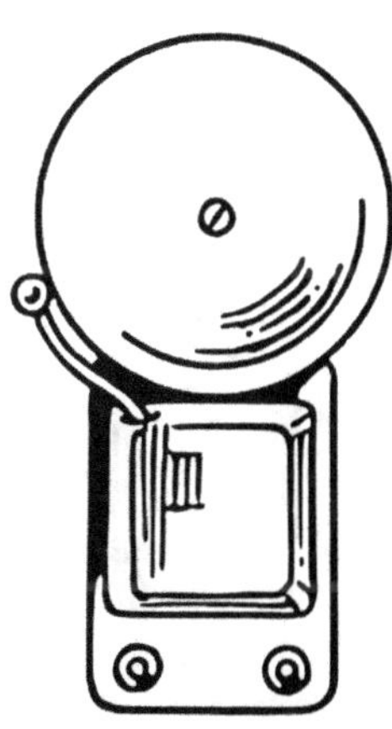

Fire Alarm Log Book

YEAR   MONTH

DATE   TIME

LOCATION   CALL NO.

CHECKS DONE

ACTION REQUIRED

ACTION TAKEN

DATE ACTION WAS LOGGED   LOGGED BY

DATE ACTION WAS CLOSED   CLOSED BY

NOTES

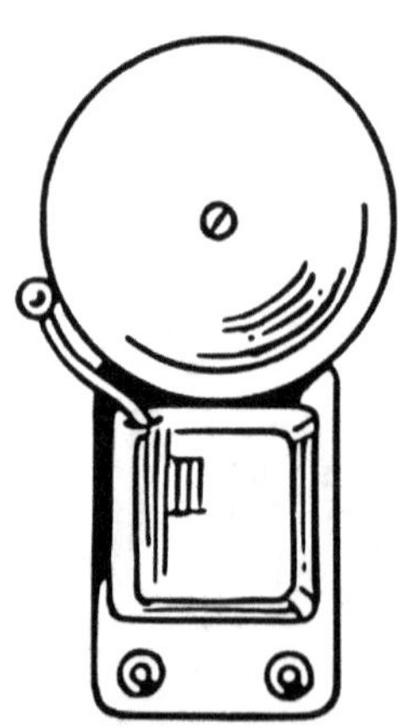

Fire Alarm Log Book

YEAR   MONTH

DATE   TIME

LOCATION   CALL NO.

CHECKS DONE

ACTION REQUIRED

ACTION TAKEN

DATE ACTION WAS LOGGED   LOGGED BY

DATE ACTION WAS CLOSED   CLOSED BY

NOTES

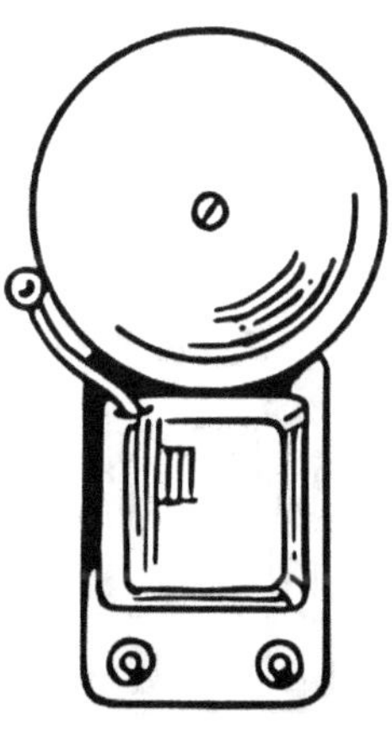

Fire Alarm Log Book

YEAR   MONTH

DATE   TIME

LOCATION   CALL NO.

CHECKS DONE

ACTION REQUIRED

ACTION TAKEN

DATE ACTION WAS LOGGED   LOGGED BY

DATE ACTION WAS CLOSED   CLOSED BY

NOTES

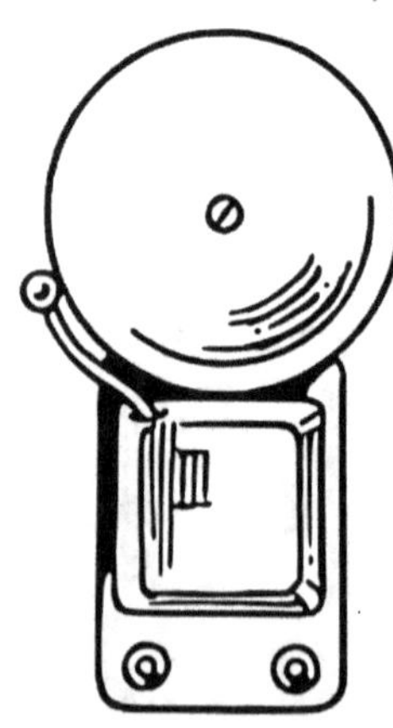

Fire Alarm Log Book

YEAR   MONTH

DATE   TIME

LOCATION   CALL NO.

CHECKS DONE

ACTION REQUIRED

ACTION TAKEN

DATE ACTION WAS LOGGED   LOGGED BY

DATE ACTION WAS CLOSED   CLOSED BY

NOTES

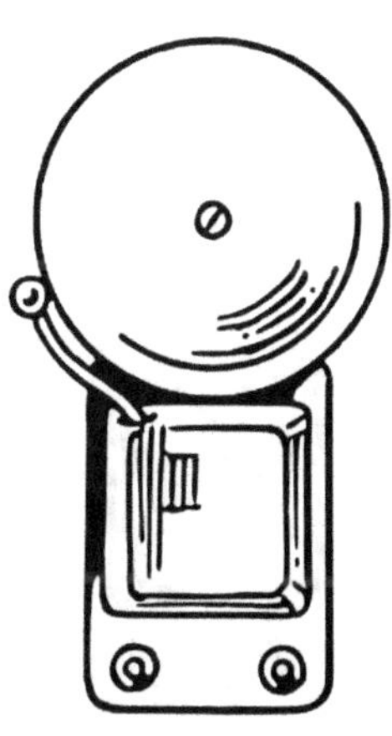

Fire Alarm Log Book

YEAR

MONTH

DATE

TIME

LOCATION

CALL NO.

CHECKS DONE

ACTION REQUIRED

ACTION TAKEN

DATE ACTION WAS LOGGED

LOGGED BY

DATE ACTION WAS CLOSED

CLOSED BY

NOTES

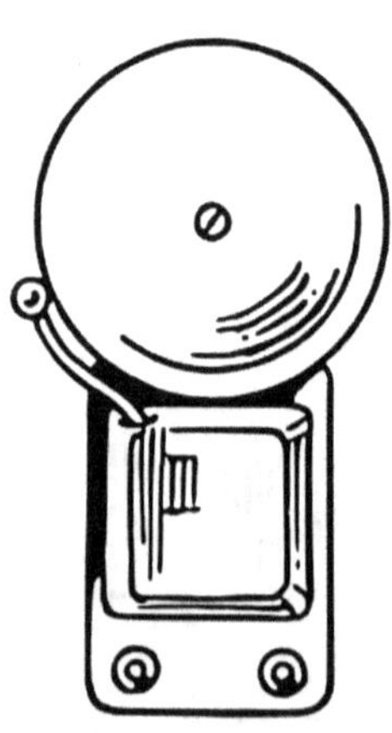

Fire Alarm Log Book

YEAR	MONTH
DATE	TIME
LOCATION	CALL NO.

CHECKS DONE

ACTION REQUIRED

ACTION TAKEN

DATE ACTION WAS LOGGED	LOGGED BY
DATE ACTION WAS CLOSED	CLOSED BY

NOTES

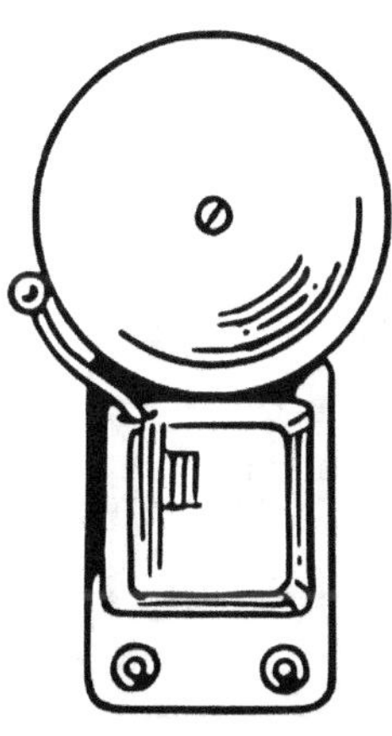

Fire Alarm Log Book

YEAR   MONTH

DATE   TIME

LOCATION   CALL NO.

CHECKS DONE

ACTION REQUIRED

ACTION TAKEN

DATE ACTION WAS LOGGED   LOGGED BY

DATE ACTION WAS CLOSED   CLOSED BY

NOTES

Fire Alarm Log Book

YEAR	**MONTH**
DATE	**TIME**
LOCATION	**CALL NO.**

CHECKS DONE

ACTION REQUIRED

ACTION TAKEN

DATE ACTION WAS LOGGED	**LOGGED BY**
DATE ACTION WAS CLOSED	**CLOSED BY**

NOTES

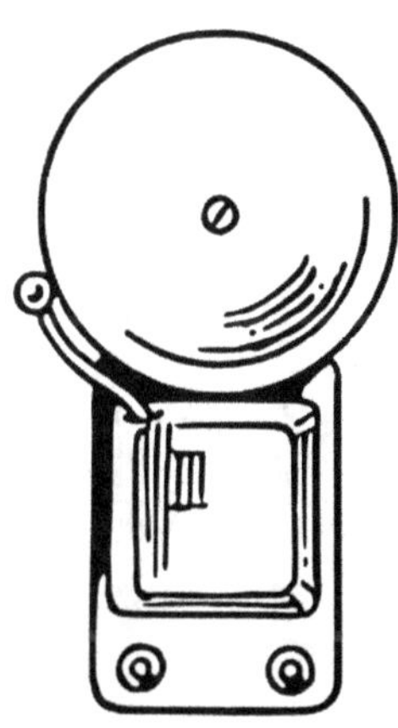

Fire Alarm Log Book

YEAR   **MONTH**

DATE   **TIME**

LOCATION   **CALL NO.**

CHECKS DONE

ACTION REQUIRED

ACTION TAKEN

DATE ACTION WAS LOGGED   **LOGGED BY**

DATE ACTION WAS CLOSED   **CLOSED BY**

NOTES

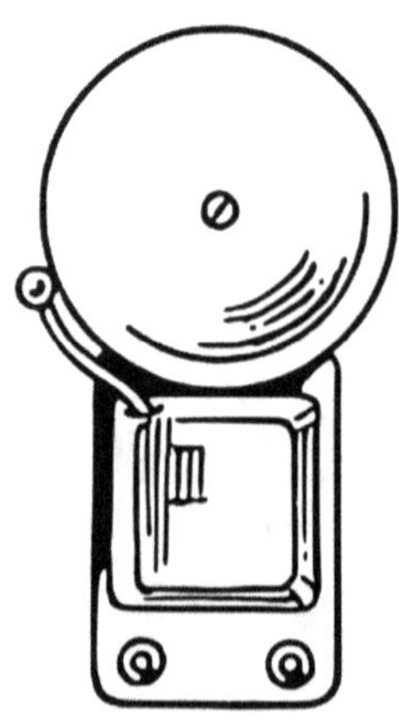

Fire Alarm Log Book

YEAR | MONTH

DATE | TIME

LOCATION | CALL NO.

CHECKS DONE

ACTION REQUIRED

ACTION TAKEN

DATE ACTION WAS LOGGED | LOGGED BY

DATE ACTION WAS CLOSED | CLOSED BY

NOTES

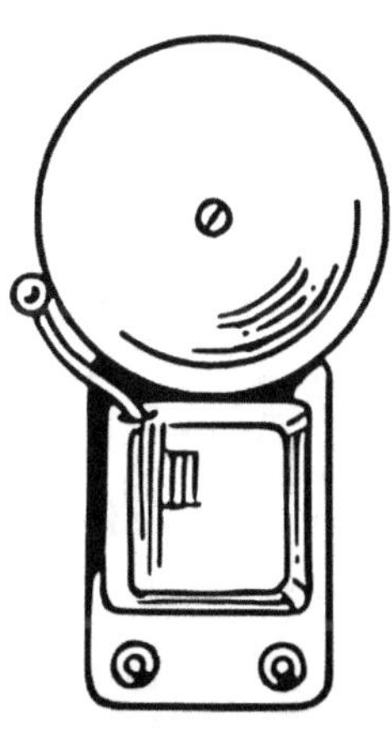

Fire Alarm Log Book

YEAR   MONTH

DATE   TIME

LOCATION   CALL NO.

CHECKS DONE

ACTION REQUIRED

ACTION TAKEN

DATE ACTION WAS LOGGED   LOGGED BY

DATE ACTION WAS CLOSED   CLOSED BY

NOTES

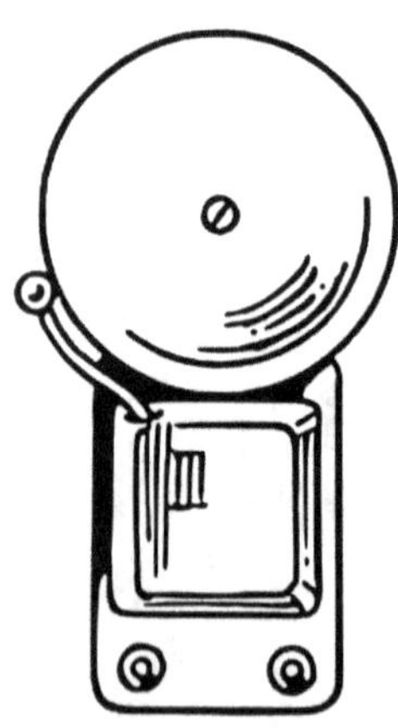

Fire Alarm Log Book

YEAR	**MONTH**
DATE	**TIME**
LOCATION	**CALL NO.**

CHECKS DONE

ACTION REQUIRED

ACTION TAKEN

DATE ACTION WAS LOGGED	**LOGGED BY**
DATE ACTION WAS CLOSED	**CLOSED BY**

NOTES

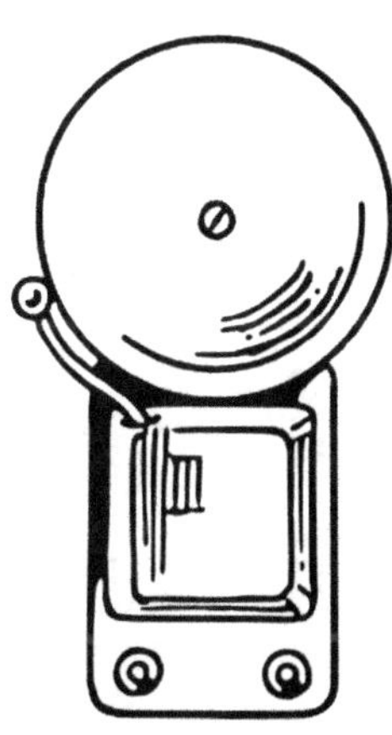

Fire Alarm Log Book

YEAR   MONTH

DATE   TIME

LOCATION   CALL NO.

CHECKS DONE

ACTION REQUIRED

ACTION TAKEN

DATE ACTION WAS LOGGED   LOGGED BY

DATE ACTION WAS CLOSED   CLOSED BY

NOTES

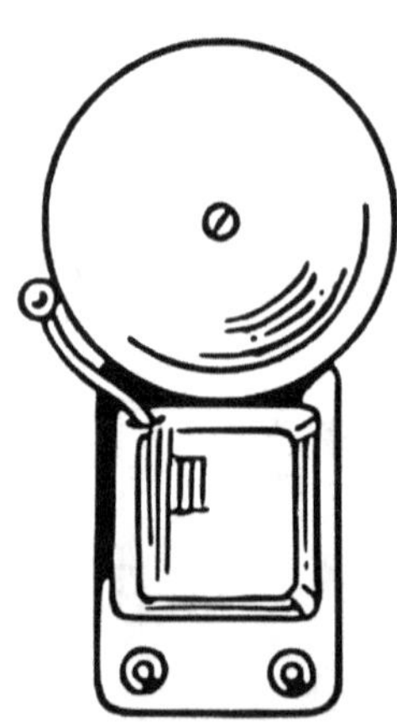

Fire Alarm Log Book

YEAR   MONTH

DATE   TIME

LOCATION   CALL NO.

CHECKS DONE

ACTION REQUIRED

ACTION TAKEN

DATE ACTION WAS LOGGED   LOGGED BY

DATE ACTION WAS CLOSED   CLOSED BY

NOTES

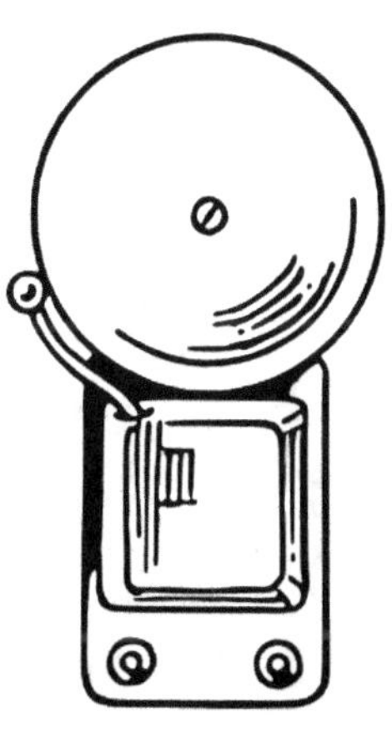

Fire Alarm Log Book

YEAR   MONTH

DATE   TIME

LOCATION   CALL NO.

CHECKS DONE

ACTION REQUIRED

ACTION TAKEN

DATE ACTION WAS LOGGED   LOGGED BY

DATE ACTION WAS CLOSED   CLOSED BY

NOTES

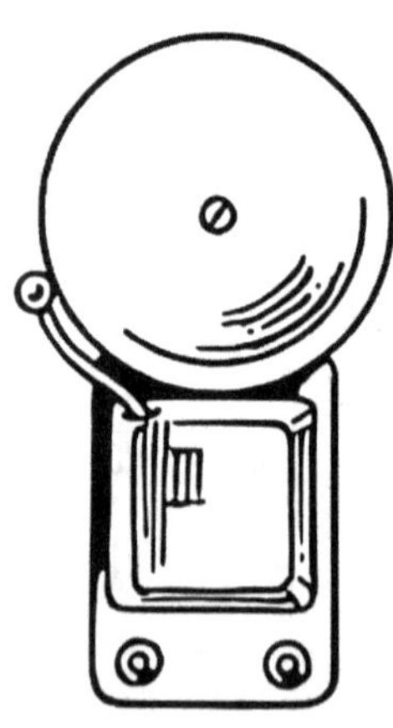

Fire Alarm Log Book

YEAR	**MONTH**
DATE	**TIME**
LOCATION	**CALL NO.**

CHECKS DONE

ACTION REQUIRED

ACTION TAKEN

DATE ACTION WAS LOGGED	**LOGGED BY**
DATE ACTION WAS CLOSED	**CLOSED BY**

NOTES

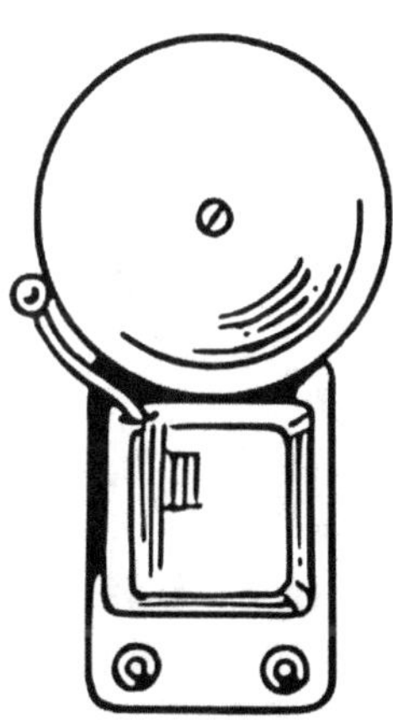

Fire Alarm Log Book

YEAR   MONTH

DATE   TIME

LOCATION   CALL NO.

CHECKS DONE

ACTION REQUIRED

ACTION TAKEN

DATE ACTION WAS LOGGED   LOGGED BY

DATE ACTION WAS CLOSED   CLOSED BY

NOTES

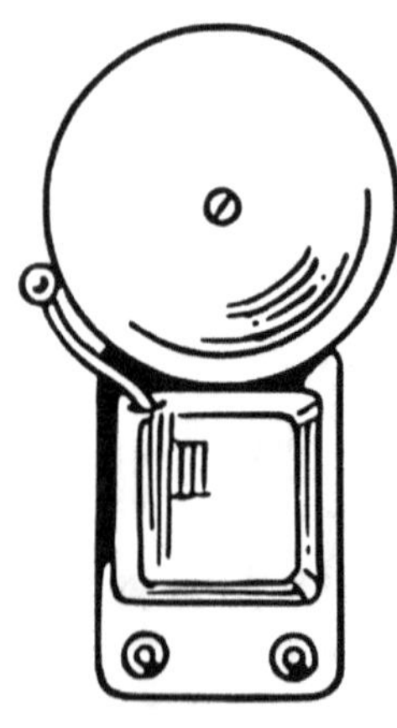

Fire Alarm Log Book

YEAR   MONTH

DATE   TIME

LOCATION   CALL NO.

CHECKS DONE

ACTION REQUIRED

ACTION TAKEN

DATE ACTION WAS LOGGED   LOGGED BY

DATE ACTION WAS CLOSED   CLOSED BY

NOTES

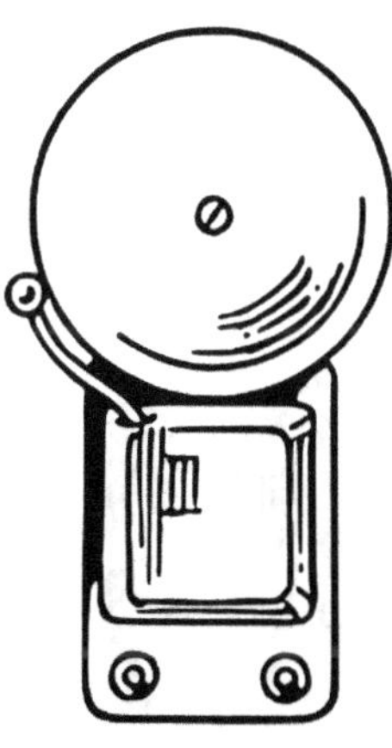

Fire Alarm Log Book

YEAR   MONTH

DATE   TIME

LOCATION   CALL NO.

CHECKS DONE

ACTION REQUIRED

ACTION TAKEN

DATE ACTION WAS LOGGED   LOGGED BY

DATE ACTION WAS CLOSED   CLOSED BY

NOTES

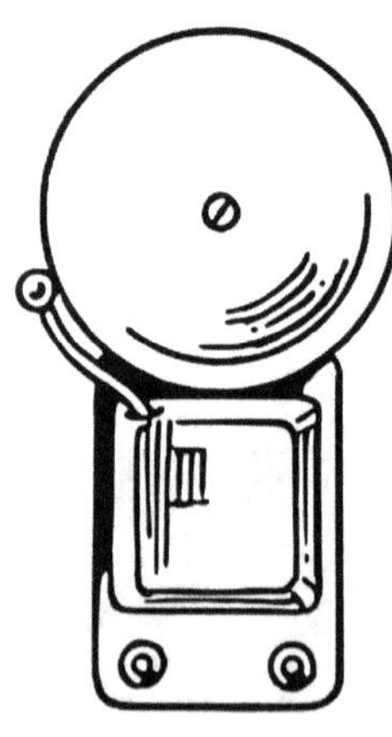

Fire Alarm Log Book

YEAR   MONTH

DATE   TIME

LOCATION   CALL NO.

CHECKS DONE

ACTION REQUIRED

ACTION TAKEN

DATE ACTION WAS LOGGED   LOGGED BY

DATE ACTION WAS CLOSED   CLOSED BY

NOTES

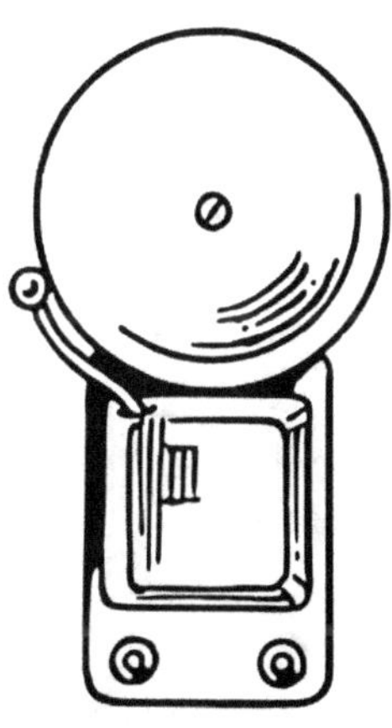

Fire Alarm Log Book

YEAR   MONTH

DATE   TIME

LOCATION   CALL NO.

CHECKS DONE

ACTION REQUIRED

ACTION TAKEN

DATE ACTION WAS LOGGED   LOGGED BY

DATE ACTION WAS CLOSED   CLOSED BY

NOTES

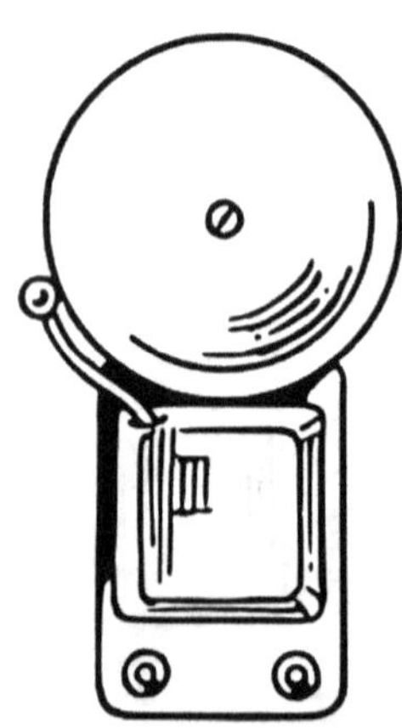

Fire Alarm Log Book

YEAR   MONTH

DATE   TIME

LOCATION   CALL NO.

CHECKS DONE

ACTION REQUIRED

ACTION TAKEN

DATE ACTION WAS LOGGED   LOGGED BY

DATE ACTION WAS CLOSED   CLOSED BY

NOTES

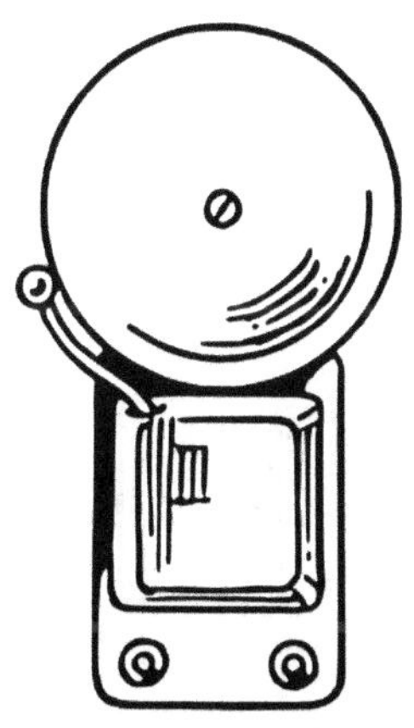

Fire Alarm Log Book

YEAR   MONTH

DATE   TIME

LOCATION   CALL NO.

CHECKS DONE

ACTION REQUIRED

ACTION TAKEN

DATE ACTION WAS LOGGED   LOGGED BY

DATE ACTION WAS CLOSED   CLOSED BY

NOTES

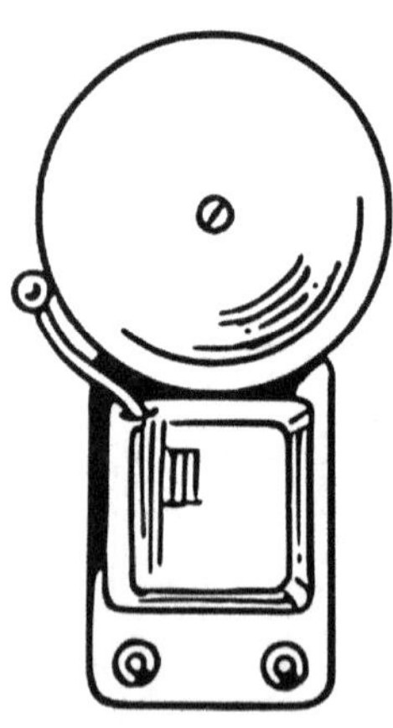

Fire Alarm Log Book

YEAR	**MONTH**
DATE	**TIME**
LOCATION	**CALL NO.**

CHECKS DONE

ACTION REQUIRED

ACTION TAKEN

DATE ACTION WAS LOGGED	**LOGGED BY**
DATE ACTION WAS CLOSED	**CLOSED BY**

NOTES

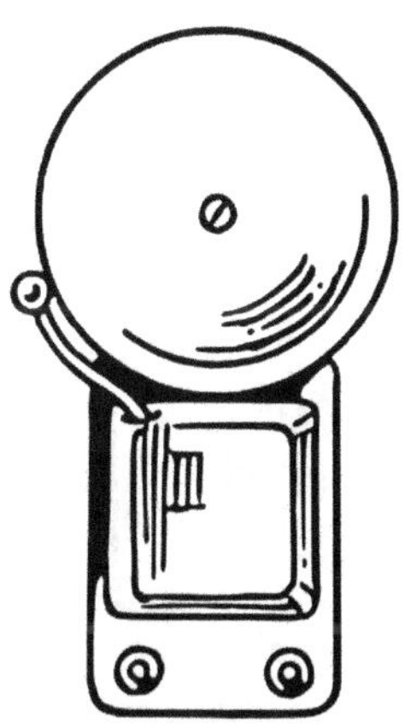

Fire Alarm Log Book

YEAR   MONTH

DATE   TIME

LOCATION   CALL NO.

CHECKS DONE

ACTION REQUIRED

ACTION TAKEN

DATE ACTION WAS LOGGED   LOGGED BY

DATE ACTION WAS CLOSED   CLOSED BY

NOTES

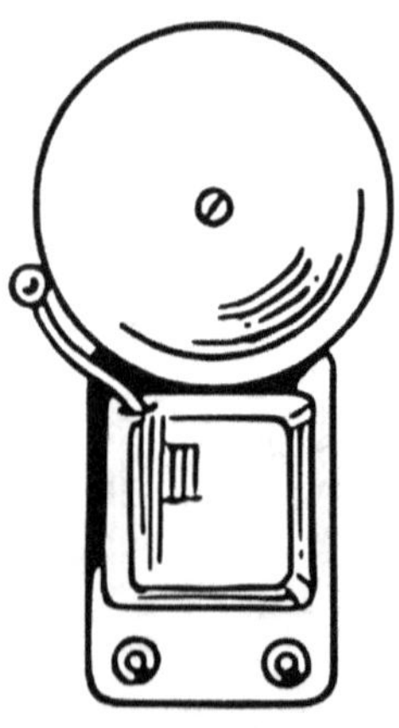

Fire Alarm Log Book

YEAR ____________________ MONTH ____________________

DATE ____________________ TIME ____________________

LOCATION ____________________ CALL NO. ____________________

CHECKS DONE

ACTION REQUIRED

ACTION TAKEN

DATE ACTION WAS LOGGED ____________________ LOGGED BY ____________________

DATE ACTION WAS CLOSED ____________________ CLOSED BY ____________________

NOTES

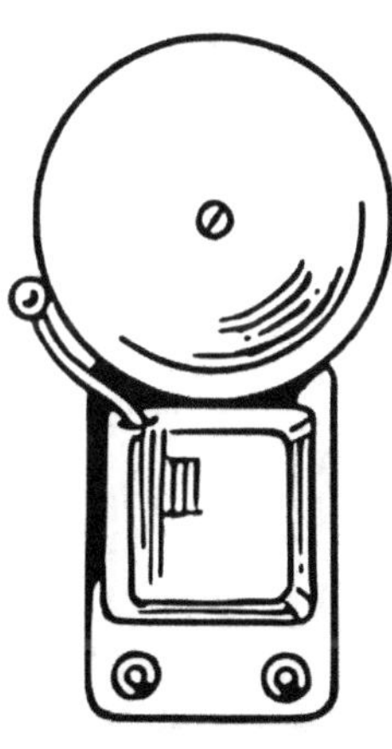

Fire Alarm Log Book

YEAR   MONTH

DATE   TIME

LOCATION   CALL NO.

CHECKS DONE

ACTION REQUIRED

ACTION TAKEN

DATE ACTION WAS LOGGED   LOGGED BY

DATE ACTION WAS CLOSED   CLOSED BY

NOTES

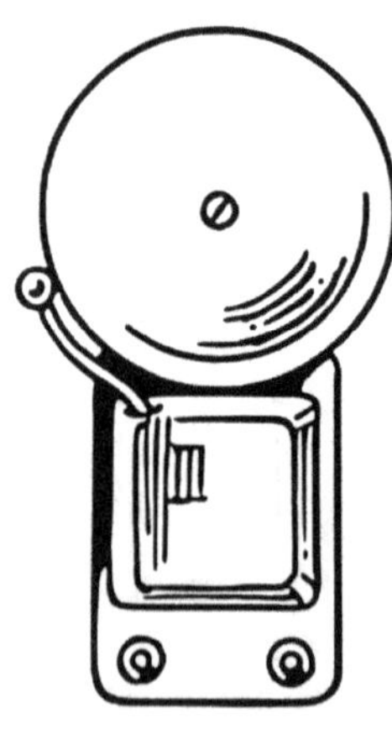

Fire Alarm Log Book

YEAR	**MONTH**
DATE	**TIME**
LOCATION	**CALL NO.**

CHECKS DONE

ACTION REQUIRED

ACTION TAKEN

DATE ACTION WAS LOGGED	**LOGGED BY**
DATE ACTION WAS CLOSED	**CLOSED BY**

NOTES

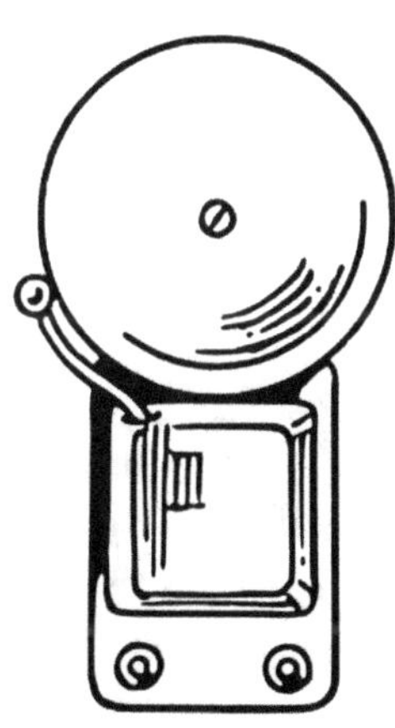

Fire Alarm Log Book

YEAR   MONTH

DATE   TIME

LOCATION   CALL NO.

CHECKS DONE

ACTION REQUIRED

ACTION TAKEN

DATE ACTION WAS LOGGED   LOGGED BY

DATE ACTION WAS CLOSED   CLOSED BY

NOTES

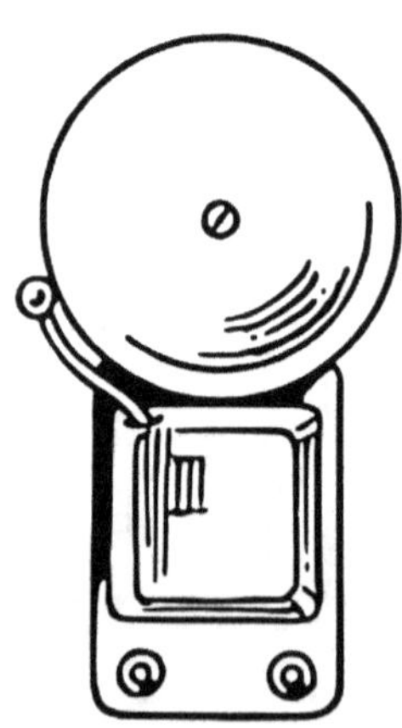

Fire Alarm Log Book

YEAR	MONTH
DATE	TIME
LOCATION	CALL NO.

CHECKS DONE

ACTION REQUIRED

ACTION TAKEN

DATE ACTION WAS LOGGED	LOGGED BY
DATE ACTION WAS CLOSED	CLOSED BY

NOTES

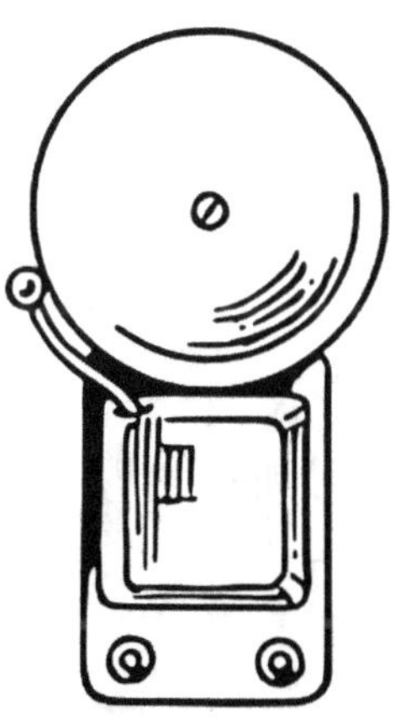

Fire Alarm Log Book

YEAR   MONTH

DATE   TIME

LOCATION   CALL NO.

CHECKS DONE

ACTION REQUIRED

ACTION TAKEN

DATE ACTION WAS LOGGED   LOGGED BY

DATE ACTION WAS CLOSED   CLOSED BY

NOTES

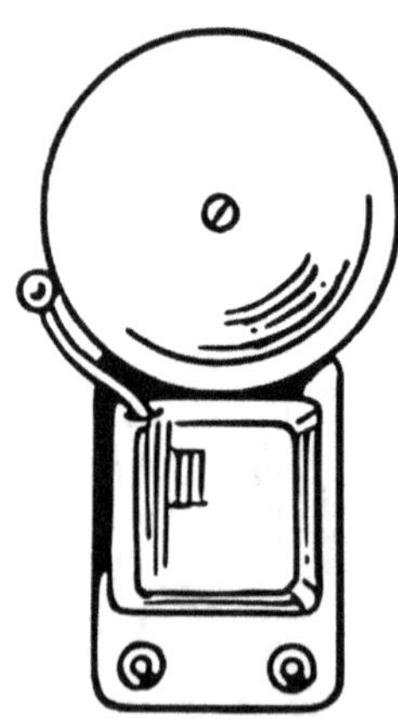

Fire Alarm Log Book

YEAR	**MONTH**	
DATE	**TIME**	
LOCATION	**CALL NO.**	

CHECKS DONE

ACTION REQUIRED

ACTION TAKEN

DATE ACTION WAS LOGGED	**LOGGED BY**
DATE ACTION WAS CLOSED	**CLOSED BY**

NOTES

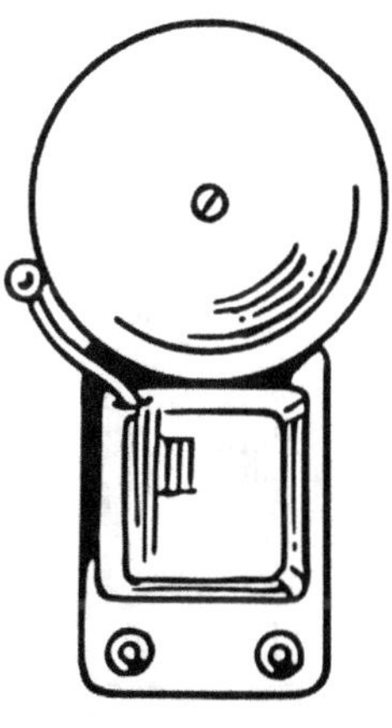

Fire Alarm Log Book

YEAR	**MONTH**
DATE	**TIME**
LOCATION	**CALL NO.**

CHECKS DONE

ACTION REQUIRED

ACTION TAKEN

DATE ACTION WAS LOGGED	**LOGGED BY**
DATE ACTION WAS CLOSED	**CLOSED BY**

NOTES

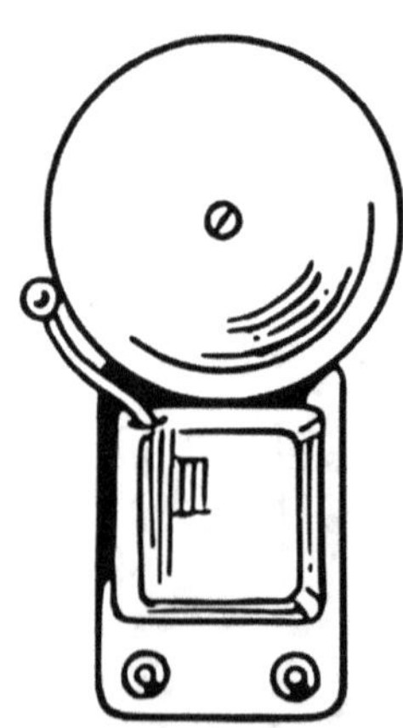

Fire Alarm Log Book

YEAR   **MONTH**

DATE   **TIME**

LOCATION   **CALL NO.**

CHECKS DONE

ACTION REQUIRED

ACTION TAKEN

DATE ACTION WAS LOGGED   **LOGGED BY**

DATE ACTION WAS CLOSED   **CLOSED BY**

NOTES

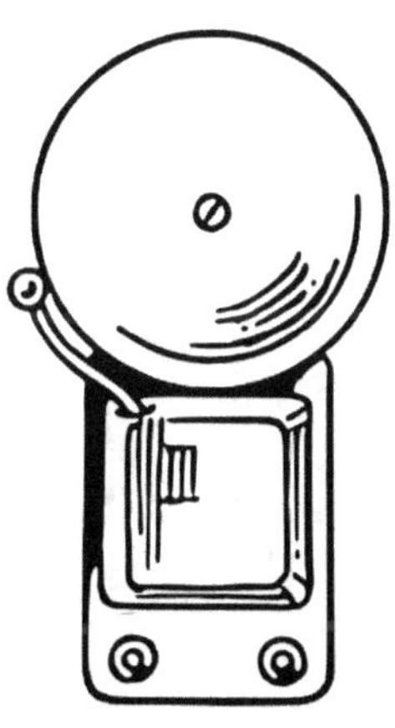

Fire Alarm Log Book

YEAR   MONTH

DATE   TIME

LOCATION   CALL NO.

CHECKS DONE

ACTION REQUIRED

ACTION TAKEN

DATE ACTION WAS LOGGED   LOGGED BY

DATE ACTION WAS CLOSED   CLOSED BY

NOTES

Fire Alarm Log Book

YEAR   MONTH

DATE   TIME

LOCATION   CALL NO.

CHECKS DONE

ACTION REQUIRED

ACTION TAKEN

DATE ACTION WAS LOGGED   LOGGED BY

DATE ACTION WAS CLOSED   CLOSED BY

NOTES

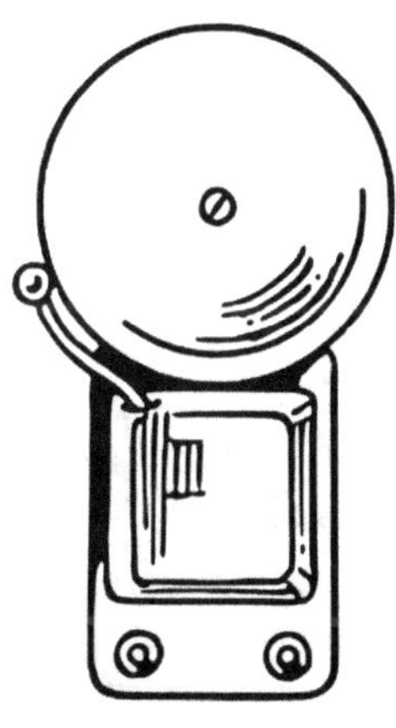

Fire Alarm Log Book

YEAR	**MONTH**
DATE	**TIME**
LOCATION	**CALL NO.**

CHECKS DONE

ACTION REQUIRED

ACTION TAKEN

DATE ACTION WAS LOGGED	**LOGGED BY**
DATE ACTION WAS CLOSED	**CLOSED BY**

NOTES

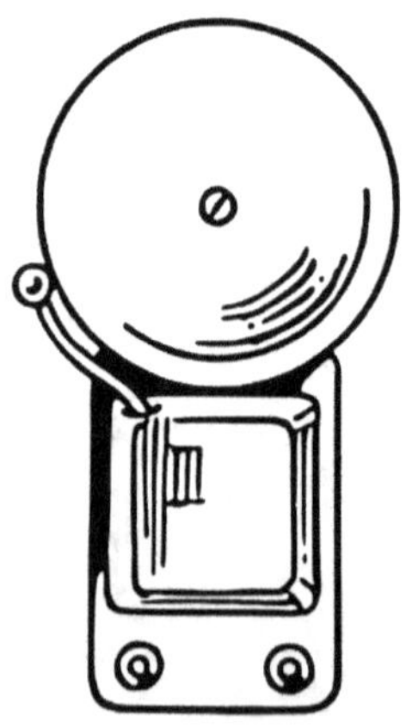

Fire Alarm Log Book

YEAR	**MONTH**
DATE	**TIME**
LOCATION	**CALL NO.**

CHECKS DONE

ACTION REQUIRED

ACTION TAKEN

DATE ACTION WAS LOGGED	**LOGGED BY**
DATE ACTION WAS CLOSED	**CLOSED BY**

NOTES

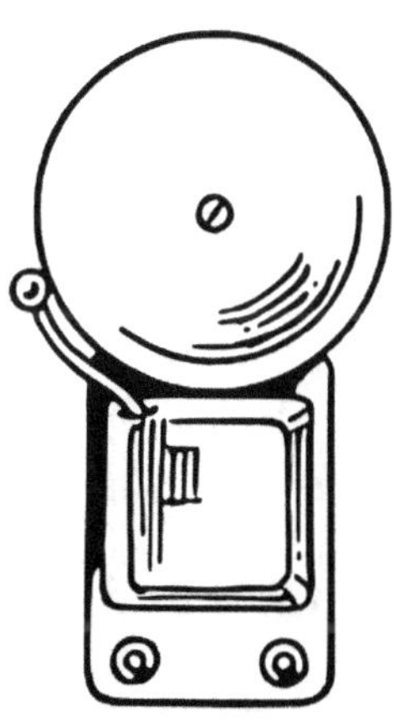

Fire Alarm Log Book

YEAR   MONTH

DATE   TIME

LOCATION   CALL NO.

CHECKS DONE

ACTION REQUIRED

ACTION TAKEN

DATE ACTION WAS LOGGED   LOGGED BY

DATE ACTION WAS CLOSED   CLOSED BY

NOTES

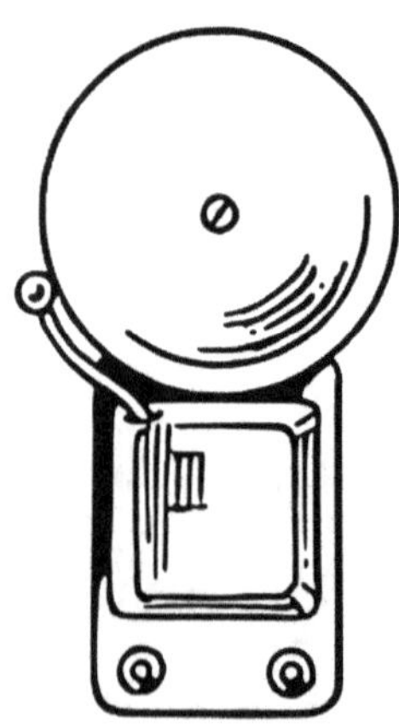

Fire Alarm Log Book

YEAR

MONTH

DATE

TIME

LOCATION

CALL NO.

CHECKS DONE

ACTION REQUIRED

ACTION TAKEN

DATE ACTION WAS LOGGED

LOGGED BY

DATE ACTION WAS CLOSED

CLOSED BY

NOTES

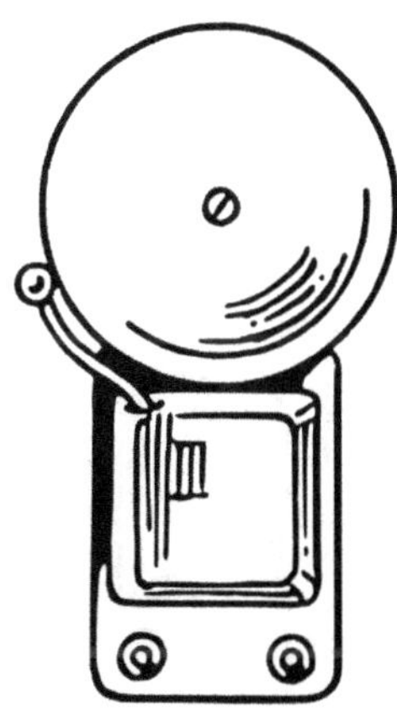

Fire Alarm Log Book

YEAR   MONTH

DATE   TIME

LOCATION   CALL NO.

CHECKS DONE

ACTION REQUIRED

ACTION TAKEN

DATE ACTION WAS LOGGED   LOGGED BY

DATE ACTION WAS CLOSED   CLOSED BY

NOTES

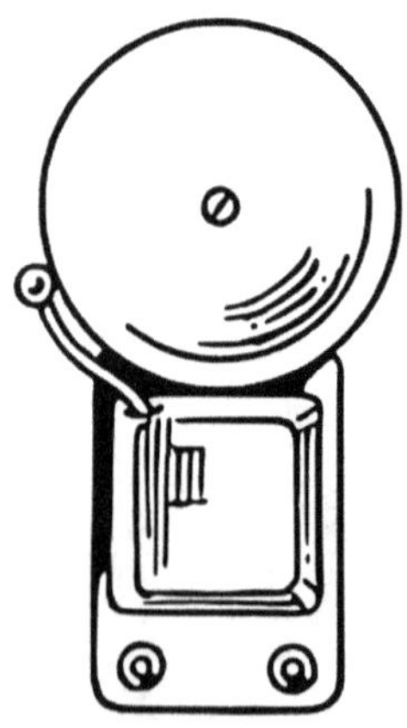

Fire Alarm Log Book

YEAR   MONTH

DATE   TIME

LOCATION   CALL NO.

CHECKS DONE

ACTION REQUIRED

ACTION TAKEN

DATE ACTION WAS LOGGED   LOGGED BY

DATE ACTION WAS CLOSED   CLOSED BY

NOTES

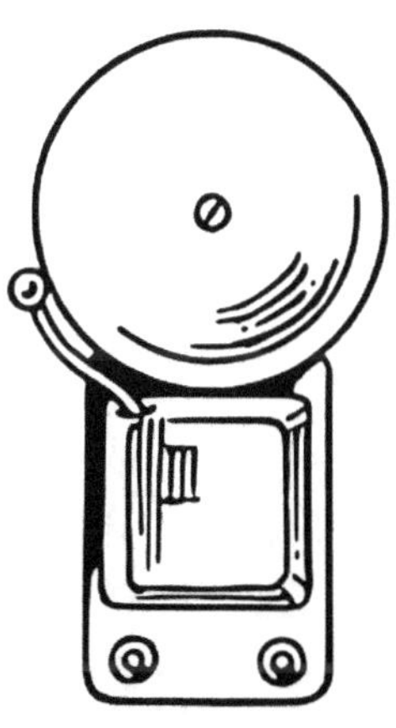

Fire Alarm Log Book

YEAR   MONTH

DATE   TIME

LOCATION   CALL NO.

CHECKS DONE

ACTION REQUIRED

ACTION TAKEN

DATE ACTION WAS LOGGED   LOGGED BY

DATE ACTION WAS CLOSED   CLOSED BY

NOTES

Fire Alarm Log Book

YEAR   MONTH

DATE   TIME

LOCATION   CALL NO.

CHECKS DONE

ACTION REQUIRED

ACTION TAKEN

DATE ACTION WAS LOGGED   LOGGED BY

DATE ACTION WAS CLOSED   CLOSED BY

NOTES

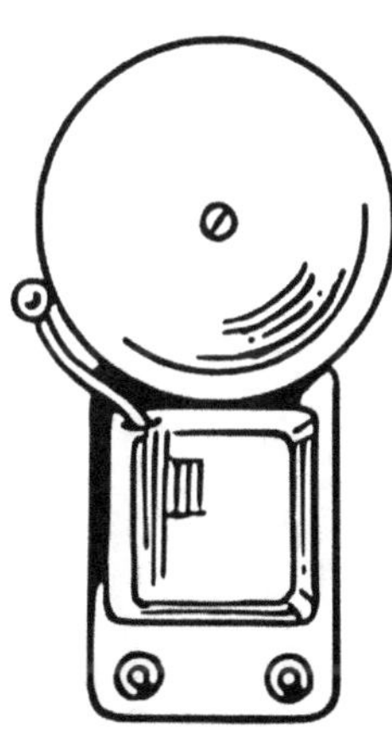

Fire Alarm Log Book

YEAR	**MONTH**
DATE	**TIME**
LOCATION	**CALL NO.**

CHECKS DONE

ACTION REQUIRED

ACTION TAKEN

DATE ACTION WAS LOGGED	**LOGGED BY**
DATE ACTION WAS CLOSED	**CLOSED BY**

NOTES

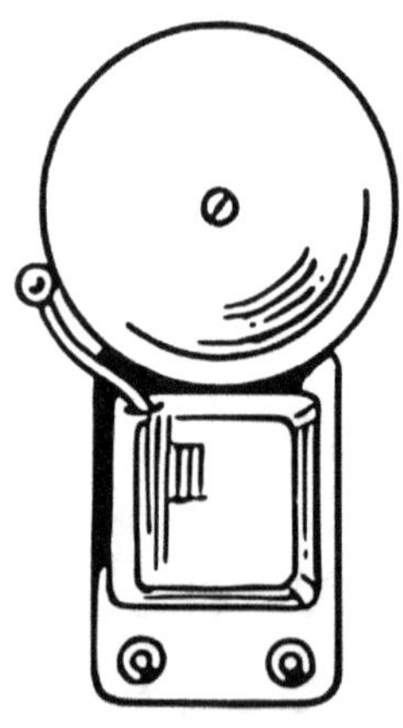

Fire Alarm Log Book

YEAR   **MONTH**

DATE   **TIME**

LOCATION   **CALL NO.**

CHECKS DONE

ACTION REQUIRED

ACTION TAKEN

DATE ACTION WAS LOGGED   **LOGGED BY**

DATE ACTION WAS CLOSED   **CLOSED BY**

NOTES

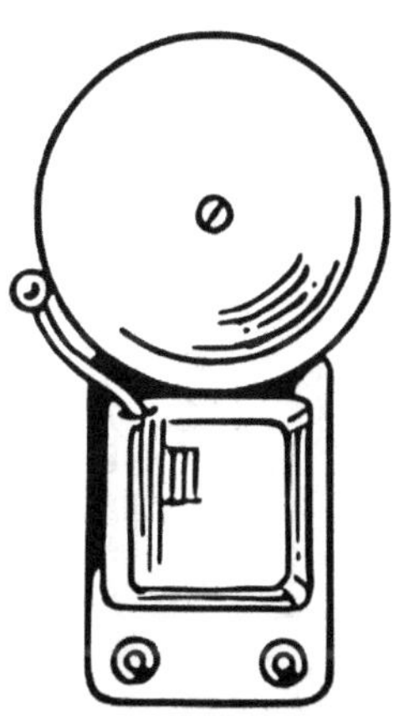

Fire Alarm Log Book

YEAR   MONTH

DATE   TIME

LOCATION   CALL NO.

CHECKS DONE

ACTION REQUIRED

ACTION TAKEN

DATE ACTION WAS LOGGED   LOGGED BY

DATE ACTION WAS CLOSED   CLOSED BY

NOTES

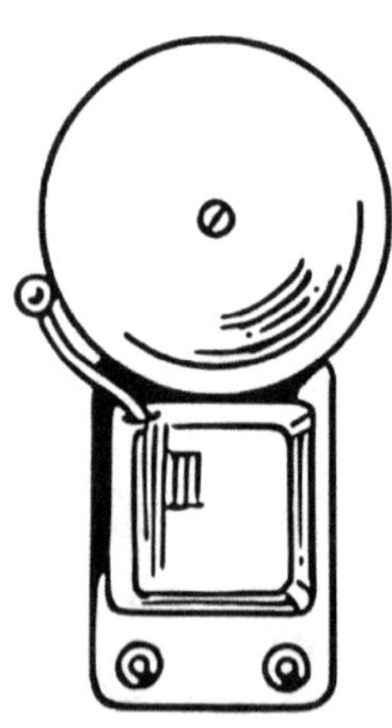

Fire Alarm Log Book 

YEAR

MONTH

DATE

TIME

LOCATION

CALL NO.

CHECKS DONE

ACTION REQUIRED

ACTION TAKEN

DATE ACTION WAS LOGGED

LOGGED BY

DATE ACTION WAS CLOSED

CLOSED BY

NOTES

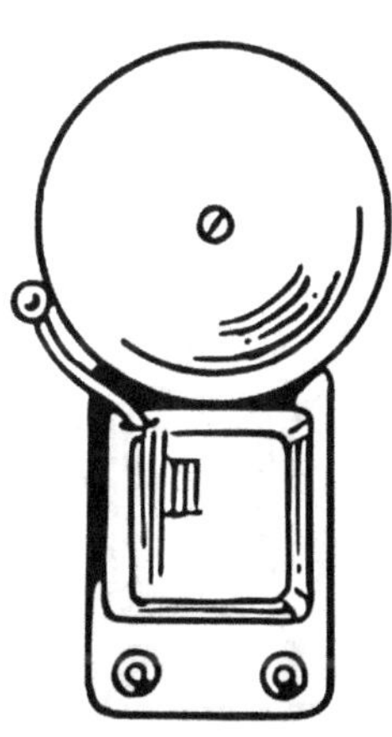

Fire Alarm Log Book

YEAR	**MONTH**
DATE	**TIME**
LOCATION	**CALL NO.**

CHECKS DONE

ACTION REQUIRED

ACTION TAKEN

DATE ACTION WAS LOGGED	**LOGGED BY**
DATE ACTION WAS CLOSED	**CLOSED BY**

NOTES

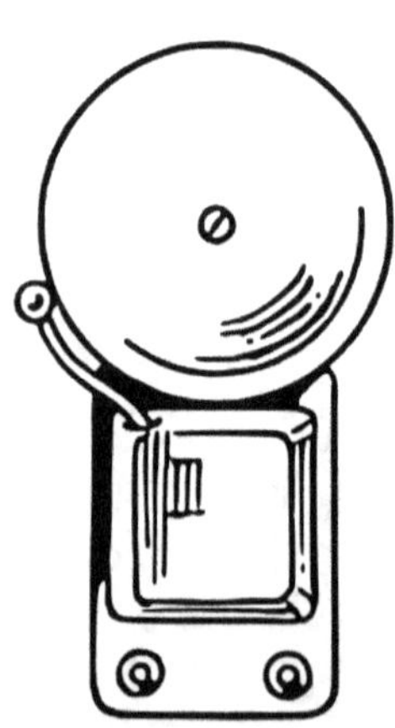

Fire Alarm Log Book

YEAR   MONTH

DATE   TIME

LOCATION   CALL NO.

CHECKS DONE

ACTION REQUIRED

ACTION TAKEN

DATE ACTION WAS LOGGED   LOGGED BY

DATE ACTION WAS CLOSED   CLOSED BY

NOTES

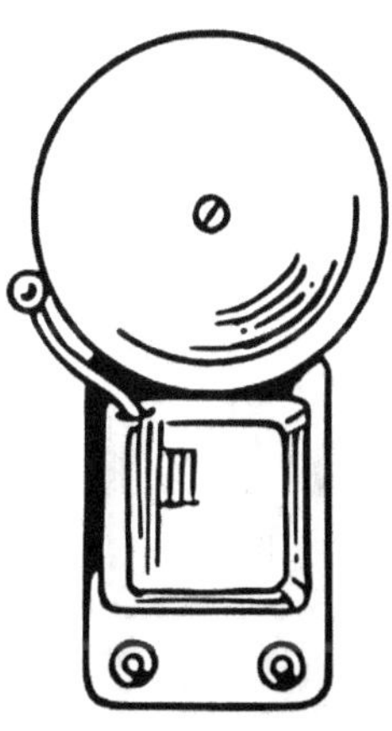

Fire Alarm Log Book

YEAR　　　　　　　　　　**MONTH**

DATE　　　　　　　　　　**TIME**

LOCATION　　　　　　　　**CALL NO.**

CHECKS DONE

ACTION REQUIRED

ACTION TAKEN

DATE ACTION WAS LOGGED　　　　　　　　**LOGGED BY**

DATE ACTION WAS CLOSED　　　　　　　　**CLOSED BY**

NOTES

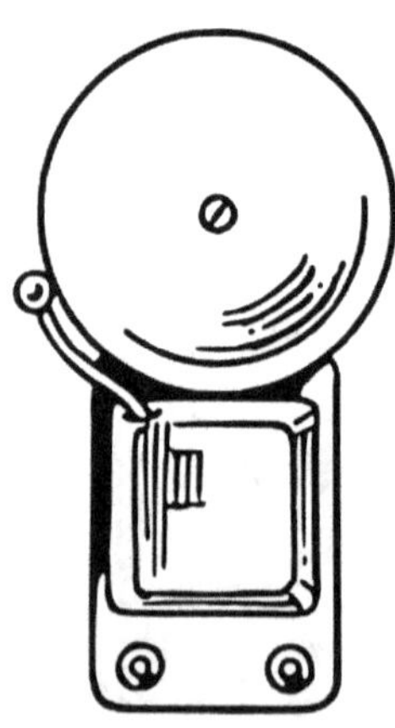

Fire Alarm Log Book

YEAR

MONTH

DATE

TIME

LOCATION

CALL NO.

CHECKS DONE

ACTION REQUIRED

ACTION TAKEN

DATE ACTION WAS LOGGED

LOGGED BY

DATE ACTION WAS CLOSED

CLOSED BY

NOTES

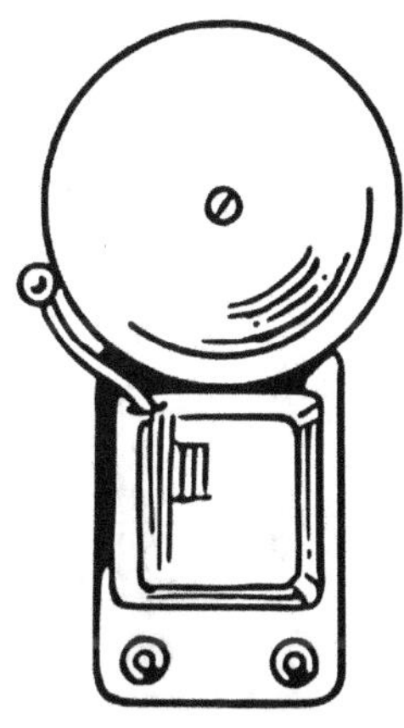

Fire Alarm Log Book

YEAR	**MONTH**
DATE	**TIME**
LOCATION	**CALL NO.**

CHECKS DONE

ACTION REQUIRED

ACTION TAKEN

DATE ACTION WAS LOGGED	**LOGGED BY**
DATE ACTION WAS CLOSED	**CLOSED BY**

NOTES

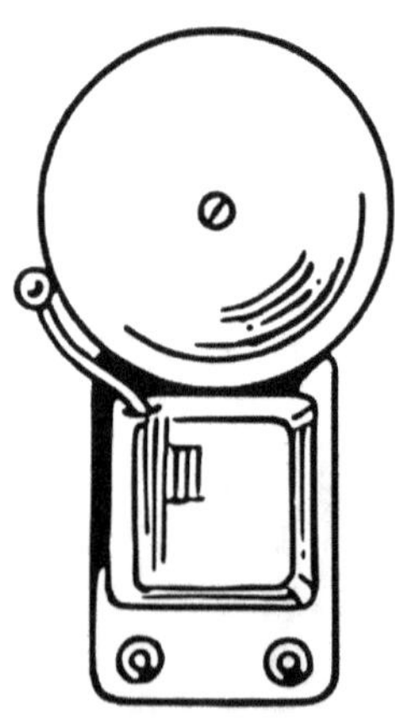

Fire Alarm Log Book

YEAR	**MONTH**
DATE	**TIME**
LOCATION	**CALL NO.**

CHECKS DONE

ACTION REQUIRED

ACTION TAKEN

DATE ACTION WAS LOGGED	**LOGGED BY**
DATE ACTION WAS CLOSED	**CLOSED BY**

NOTES

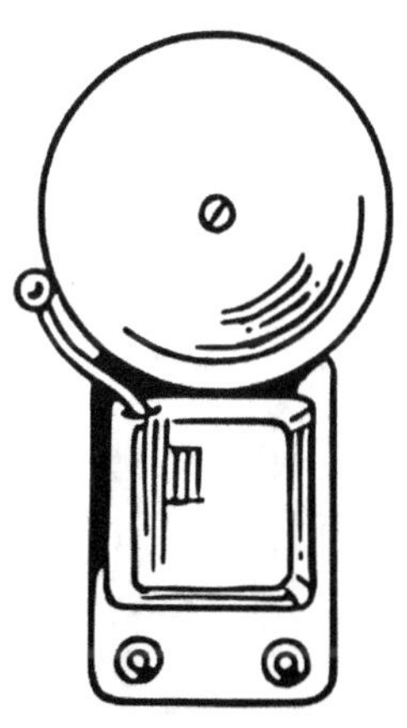

Fire Alarm Log Book

YEAR   MONTH

DATE   TIME

LOCATION   CALL NO.

CHECKS DONE

ACTION REQUIRED

ACTION TAKEN

DATE ACTION WAS LOGGED   LOGGED BY

DATE ACTION WAS CLOSED   CLOSED BY

NOTES

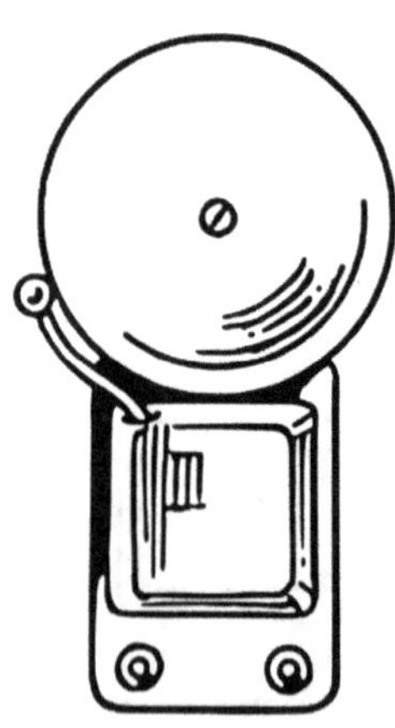

Fire Alarm Log Book

YEAR   MONTH

DATE   TIME

LOCATION   CALL NO.

CHECKS DONE

ACTION REQUIRED

ACTION TAKEN

DATE ACTION WAS LOGGED   LOGGED BY

DATE ACTION WAS CLOSED   CLOSED BY

NOTES

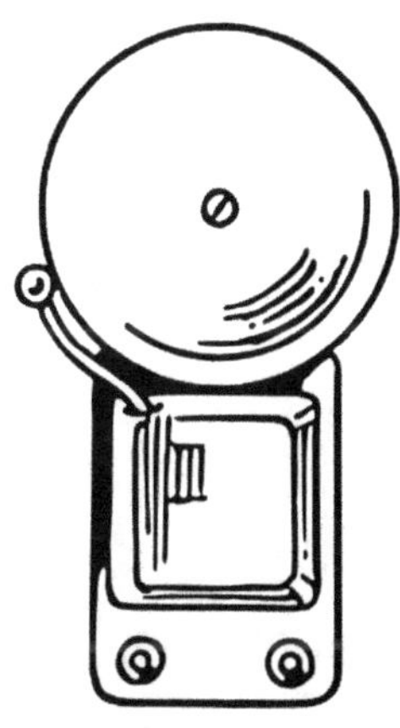

Fire Alarm Log Book 

YEAR

MONTH

DATE

TIME

LOCATION

CALL NO.

CHECKS DONE

ACTION REQUIRED

ACTION TAKEN

DATE ACTION WAS LOGGED

LOGGED BY

DATE ACTION WAS CLOSED

CLOSED BY

NOTES

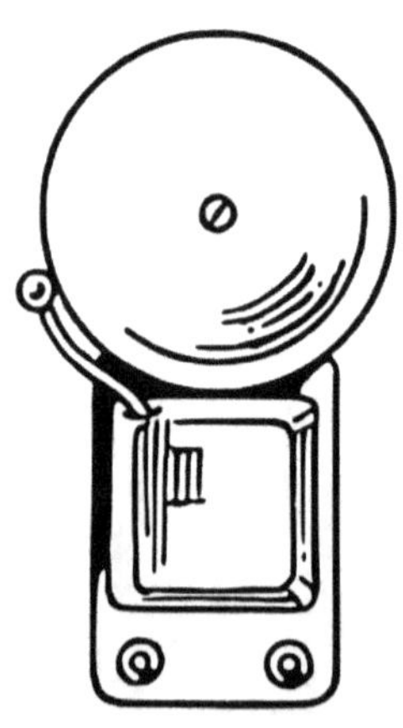

Fire Alarm Log Book

YEAR

MONTH

DATE

TIME

LOCATION

CALL NO.

CHECKS DONE

ACTION REQUIRED

ACTION TAKEN

DATE ACTION WAS LOGGED

LOGGED BY

DATE ACTION WAS CLOSED

CLOSED BY

NOTES

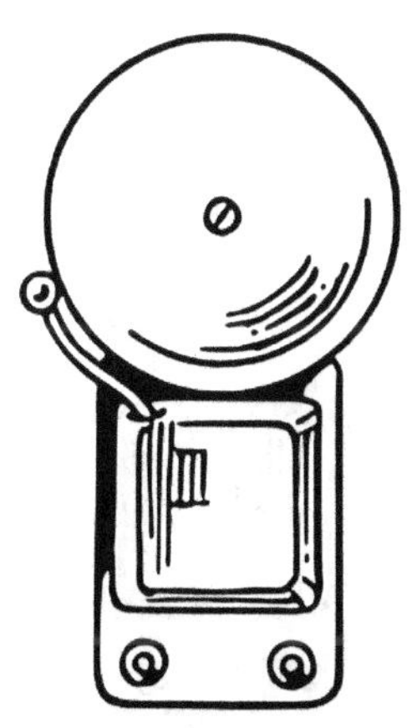

Fire Alarm Log Book

YEAR

MONTH

DATE

TIME

LOCATION

CALL NO.

CHECKS DONE

ACTION REQUIRED

ACTION TAKEN

DATE ACTION WAS LOGGED

LOGGED BY

DATE ACTION WAS CLOSED

CLOSED BY

NOTES

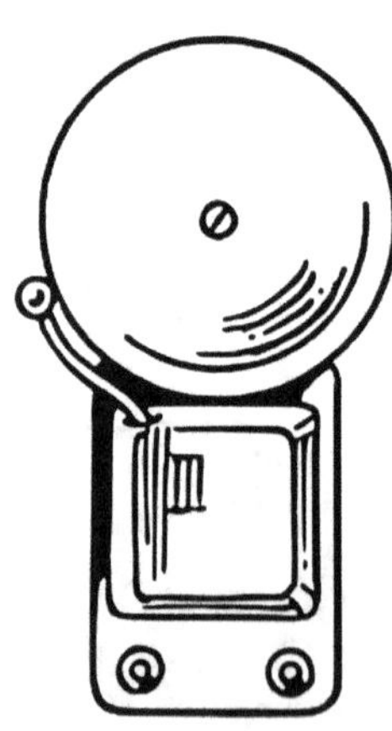

Fire Alarm Log Book

YEAR

MONTH

DATE

TIME

LOCATION

CALL NO.

CHECKS DONE

ACTION REQUIRED

ACTION TAKEN

DATE ACTION WAS LOGGED

LOGGED BY

DATE ACTION WAS CLOSED

CLOSED BY

NOTES

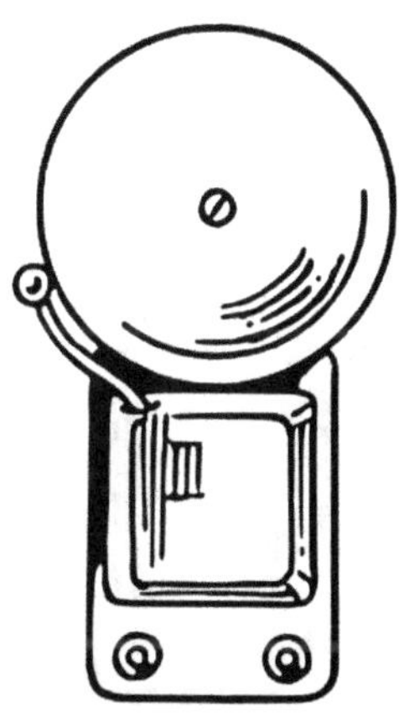

Fire Alarm Log Book

YEAR   MONTH

DATE   TIME

LOCATION   CALL NO.

CHECKS DONE

ACTION REQUIRED

ACTION TAKEN

DATE ACTION WAS LOGGED   LOGGED BY

DATE ACTION WAS CLOSED   CLOSED BY

NOTES

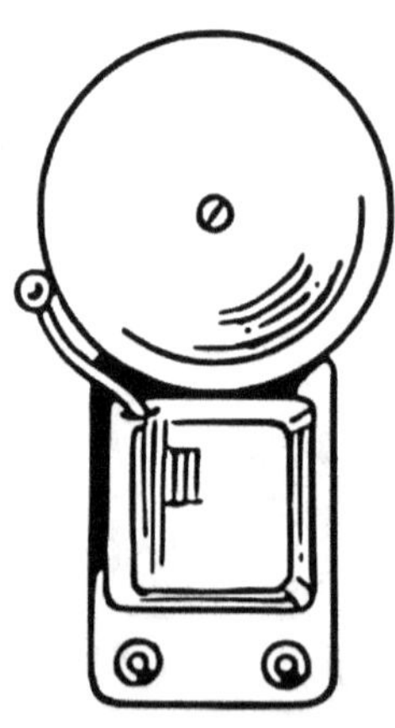

Fire Alarm Log Book

YEAR	**MONTH**
DATE	**TIME**
LOCATION	**CALL NO.**

CHECKS DONE

ACTION REQUIRED

ACTION TAKEN

DATE ACTION WAS LOGGED	**LOGGED BY**
DATE ACTION WAS CLOSED	**CLOSED BY**

NOTES

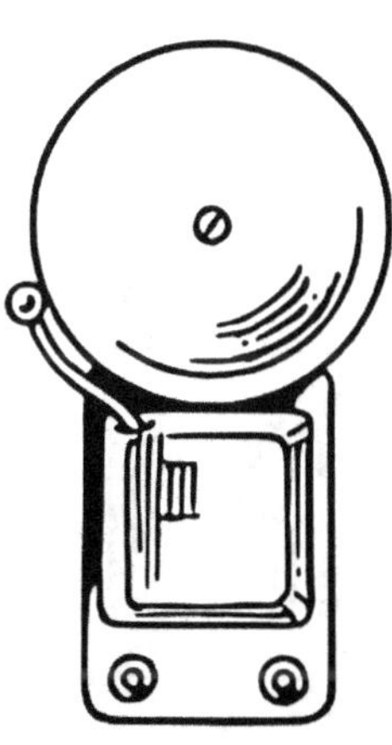

Fire Alarm Log Book

YEAR   MONTH

DATE   TIME

LOCATION   CALL NO.

CHECKS DONE

ACTION REQUIRED

ACTION TAKEN

DATE ACTION WAS LOGGED   LOGGED BY

DATE ACTION WAS CLOSED   CLOSED BY

NOTES

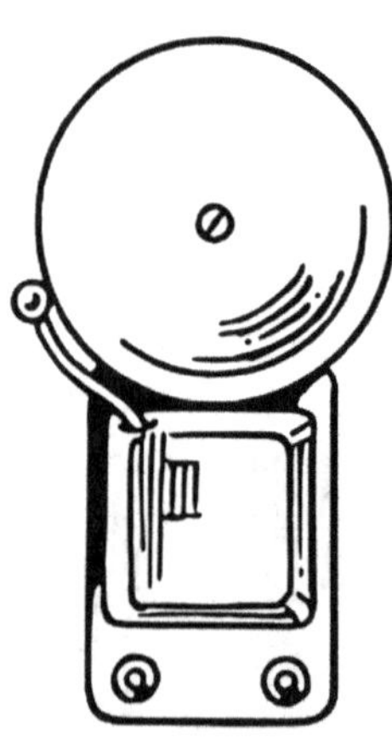

Fire Alarm Log Book

YEAR

MONTH

DATE

TIME

LOCATION

CALL NO.

CHECKS DONE

ACTION REQUIRED

ACTION TAKEN

DATE ACTION WAS LOGGED

LOGGED BY

DATE ACTION WAS CLOSED

CLOSED BY

NOTES

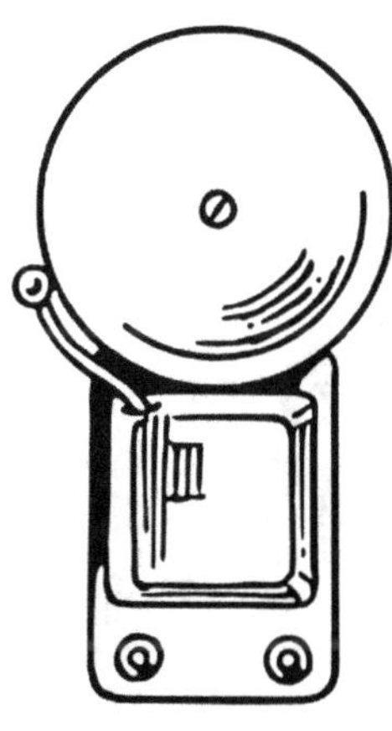

Fire Alarm Log Book

YEAR   MONTH

DATE   TIME

LOCATION   CALL NO.

CHECKS DONE

ACTION REQUIRED

ACTION TAKEN

DATE ACTION WAS LOGGED   LOGGED BY

DATE ACTION WAS CLOSED   CLOSED BY

NOTES

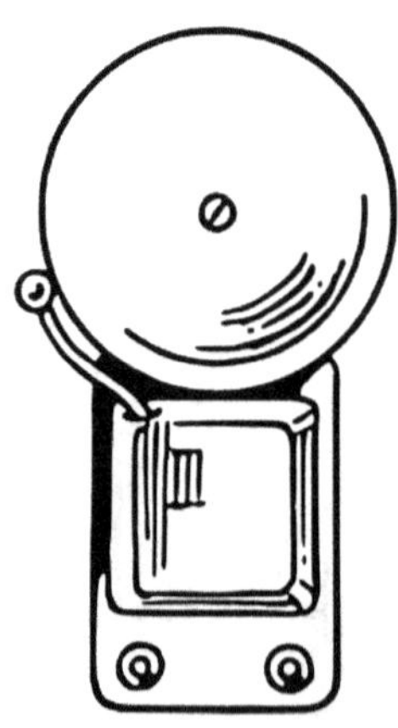

Fire Alarm Log Book

YEAR   MONTH

DATE   TIME

LOCATION   CALL NO.

CHECKS DONE

ACTION REQUIRED

ACTION TAKEN

DATE ACTION WAS LOGGED   LOGGED BY

DATE ACTION WAS CLOSED   CLOSED BY

NOTES

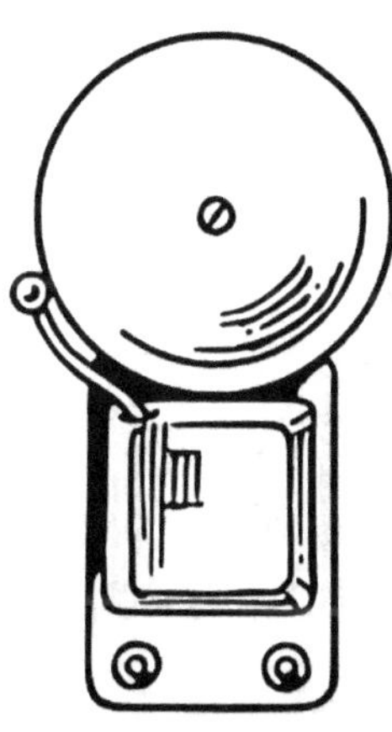

Fire Alarm Log Book

YEAR   MONTH

DATE   TIME

LOCATION   CALL NO.

CHECKS DONE

ACTION REQUIRED

ACTION TAKEN

DATE ACTION WAS LOGGED   LOGGED BY

DATE ACTION WAS CLOSED   CLOSED BY

NOTES

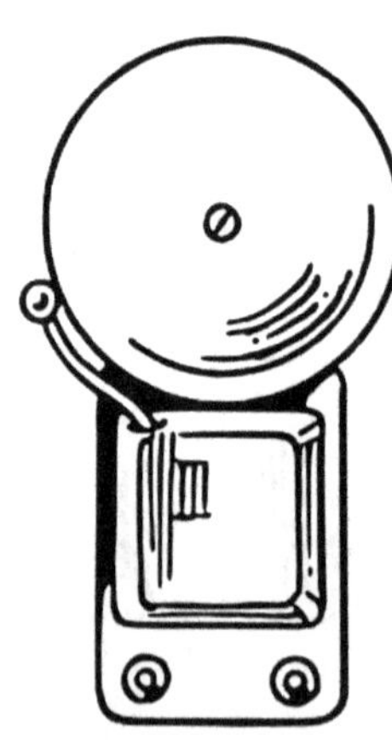

Fire Alarm Log Book

YEAR

MONTH

DATE

TIME

LOCATION

CALL NO.

CHECKS DONE

ACTION REQUIRED

ACTION TAKEN

DATE ACTION WAS LOGGED

LOGGED BY

DATE ACTION WAS CLOSED

CLOSED BY

NOTES

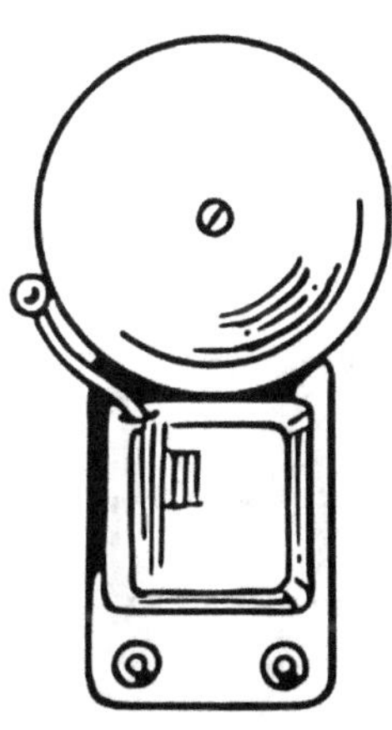

Fire Alarm Log Book

YEAR   MONTH

DATE   TIME

LOCATION   CALL NO.

CHECKS DONE

ACTION REQUIRED

ACTION TAKEN

DATE ACTION WAS LOGGED   LOGGED BY

DATE ACTION WAS CLOSED   CLOSED BY

NOTES

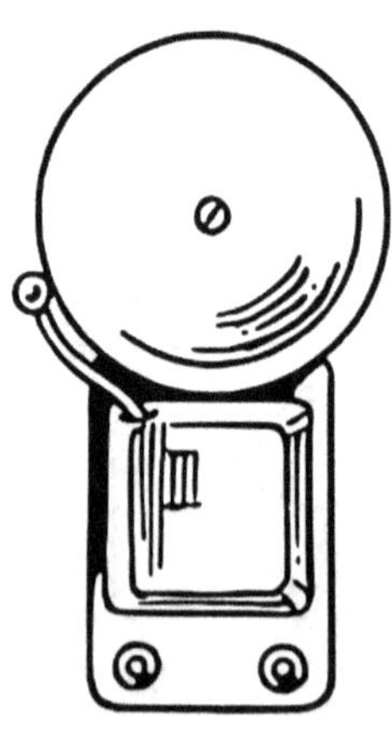